ENGINEERING MATHEMATICS

Edited by
Lalaine Suarez Figueroa

Delve Publishing

Engineering Mathematics

Edited by Lalaine Suarez Figueroa

Edition 2018

Delve Publishing

Published by
Delve Publishing LLC
708 3rd Avenue, 6th Floor
New York
NY 10801
United States of America
www.delvepublishing.com

ISBN: 978-1-68095-687-0 (Hardback)

Printed in United States of America

Library of Congress Control Number: 2016920459

About the Editor

Lalaine Suarez Figueroa

Lalaine gained her Bachelor's Degree in Instrumentation and Control Engineering from Rizal Technological University in 2007. She had gained her Certification from Philippine Instrumentation and Control Society in 2008. She is currently working as a Designer in a multinational engineering and construction firm which had an office based in the Philippines.

Table of Contents

2 FRACTIONS, DECIMAL AND PERCENTAGE

6 POTENTIAL THEORY

7 PROBABILITY

8 REAL ANALYSIS

9 FOURIER ANALYSIS

List of Abbreviations

CLT	Central Limit Theorem
DFT	Discrete Fourier Transform
FFT	Fast Fourier Transform
GCLT	Generalized Central Limit Theorem
LLN	Law of Large Numbers
SVD	Singular Value Decomposition

Preface

Engineering Mathematics is the art of applying mathematics to complex real-world problems. It combines mathematical theory, practical engineering and scientific computing to address today's technological challenges. It is a creative and exciting discipline, spanning traditional boundaries. Engineering mathematicians can be found in an extraordinarily wide range of careers, from designing next generation Formula One cars to working at the cutting edge of robotics, from running their own business creating new autonomous vehicles to developing innovative indices for leading global financial institutions. Mathematics or particularly applied mathematics is widely used in every engineering fields. In the various engineering fields, different forms of math may be applied to achieve a specific solution. For instance, linear algebra is widely applied in electrical engineering to design electrical circuits. Boolean algebra is the basis of digital logic design, a key component of building computers. In civil and mechanical engineering, geometry is used to design structures and ensure that they perform in a safe and meaningful way. When designing a bridge, civil engineers may use differential equations to calculate the approximate size of the supporting piers needed. They may also use complex mathematical modeling to predict the lifespan of the structure to be constructed. Industrial engineers use calculus to find the rate at which certain variables will change in a given operation. Industrial engineering routinely entails making approximations and safe estimations to ensure that all processes are properly controlled. One or more forms of mathematics are used in these approximations and estimations.

Content Coverage

Chapter One aims to clarify the basic arithmetic that deals with properties of the counting numbers and fractions and the basic operations applied to these numbers.

Chapter Two presents an overview of Fractions, Decimal and Percentage and their applications.

Chapter Three provides the coverage of quadratic equations used particularly in the natural sciences and engineering disciplines. It includes the study of lines, planes, and subspaces, but is also concerned with properties common to all vector spaces.

Chapter Four presents the study of linear sets of equations and their transformation properties.

Chapter Five presents an overview of mathematical models used particularly in the natural sciences and engineering disciplines but also in the social sciences; physicists, engineers, computer scientists, and economists use mathematical models most extensively.

Chapter Six focuses on potential theory that deals with the study of harmonic functions.

Chapter Seven focuses on the concepts of probability that deals with calculating the likelihood of a given event's occurrence, which is expressed as a number between 1 and 0.

Chapter Eight is concerned with studying the behavior and properties of functions, sequences, and sets on the real number line, which we denote as the mathematically familiar R.

Chapter Nine focuses on the study of the way general functions may be represented or approximated by sums of simpler trigonometric functions.

CHAPTER 1

Basic Arithmetic

Chapter Contents

Arithmetic is a branch of mathematics that deals with properties of the counting (and also whole) numbers and fractions and the basic operations applied to these numbers. As a matter of fact, as a noun in the above sense, the word is used quite seldom. In the early grades, when numbers are the main object of study, the subject is often designated as *mathematics*. The *mathematics* appellation sticks around until much later when it paradoxically becomes *Algebra* I. The latter is usually associated with the use of letters as place holders for generic or unknown numbers.

INTRODUCTION

Arithmetic or arithmetics is the oldest and most elementary branch of mathematics. It consists of the study of numbers, especially the properties of the traditional operations between them—addition, subtraction, multiplication and division. Arithmetic is an elementary part of number theory, and number theory is considered to be one of the top-level divisions of modern mathematics, along with algebra, geometry, and analysis. The terms *arithmetic* and *higher arithmetic* were used until the beginning of the 20th century as synonyms for *number theory* and are sometimes still used to refer to a wider part of number theory.

HISTORY

The prehistory of arithmetic is limited to a small number of artifacts which may indicate the conception of addition and subtraction, the best-known being the Ishango bone from central Africa, dating from somewhere between 20,000 and 18,000 BC, although its interpretation is disputed.

The earliest written records indicate the Egyptians and Babylonians used all the elementary arithmetic operations as early as 2000 BC. These artifacts do not always reveal the specific process used for solving problems, but the characteristics of the particular numeral system strongly influence the complexity of the methods. The hieroglyphic system for Egyptian numerals, like the later Roman numerals, descended from tally marks used for counting. In both cases, this origin resulted in values that used a decimal base but did not include positional notation. Complex calculations with Roman numerals required the assistance of a counting board or the Roman abacus to obtain the results.

Early number systems that included positional notation were not decimal, including the sexagesimal (base 60) system for Babylonian numerals and the vigesimal (base 20) system that defined Maya numerals. Because of this place-value concept, the ability to reuse the same digits for different values contributed to simpler and more efficient methods of calculation.

The continuous historical development of modern arithmetic starts with the Hellenistic civilization of ancient Greece, although it originated much later than the Babylonian and Egyptian examples. Prior to the works of Euclid around 300 BC, Greek studies in mathematics overlapped with philosophical and mystical beliefs. For example, Nicomachus summarized the viewpoint of the earlier Pythagorean approach to numbers, and their relationships to each other, in his *Introduction to Arithmetic*.

Greek numerals were used by Archimedes, Diophantus and others in a positional notation not very different from ours. Because the ancient Greeks lacked a symbol for zero (until the Hellenistic period), they used three separate sets of symbols. One set for the unit's place, one for the ten's place, and one for the hundred's. Then for the thousand's place they would reuse the symbols for the unit's place, and so on. Their addition algorithm was identical to ours, and their multiplication algorithm was only very slightly different. Their long division algorithm was the same, and the square root algorithm that was once taught in schoolwas known to Archimedes, who may have invented it. He preferred it to Hero's method of successive approximation because, once computed, a digit doesn't change, and the square roots of perfect squares, such as 7485696, terminate immediately as 2736. For numbers with a fractional part, such as 546.934, they used negative powers of 60 instead of negative powers of 10 for the fractional part 0.934. The ancient Chinese used a similar positional notation. Because they also lacked a symbol for zero, they had one set of symbols for the unit's place, and a second set for the ten's place. For the hundred's place they then reused the symbols for the unit's place, and so on. Their symbols were based on the ancient counting rods. It is a complicated question to determine exactly when the Chinese started calculating with positional representation, but it was definitely before 400 BC. The Bishop of Syria, Severus Sebokht (650 AD), "Indians possess a method of calculation that no word can praise enough. Their rational system of mathematics, or of their method of calculation. I mean the system using nine symbols."

Leonardo of Pisa (Fibonacci) in 1200 AD wrote in *Liber Abaci* "The method of the Indians (Modus Indoram) surpasses any known method to compute. It's a marvelous method. They do their computations using nine figures and symbol zero".

The gradual development of Hindu–Arabic numerals independently devised the place-value concept and positional notation, which combined the simpler methods for computations with a decimal base and the use of a digit representing 0. This allowed the system to consistently represent both large and small integers. This approach eventually replaced all other systems. In the early 6th century AD, the Indian mathematician Aryabhata incorporated an existing version of this system in his work, and experimented with different notations. In the 7th century, Brahmagupta established the use of 0 as a separate number and determined the results for multiplication, division, addition and subtraction of zero and all other numbers, except for the result of division by 0. His contemporary, the Syriac bishop Severus Sebokht described the excellence of this system as "... valuable methods of calculation which surpass description". The Arabs also learned this new method and called it *hesab.*

Leibniz's Stepped Reckoner was the first calculator that could perform all four arithmetic operations.

Although the Codex Vigilanus described an early form of Arabic numerals (omitting 0) by 976 AD, Fibonacci was primarily responsible for spreading their use throughout Europe after the publication of his book *Liber Abaci* in 1202. He considered the significance of this "new" representation of numbers, which he styled the "Method of the Indians" (Latin *Modus Indorum*), so fundamental that all related mathematical foundations, including the results of Pythagoras and the algorism describing the methods for performing actual calculations, were "almost a mistake" in comparison.

In the Middle Ages, arithmetic was one of the seven liberal arts taught in universities.

The flourishing of algebra in the medieval Islamic world and in Renaissance Europe was an outgrowth of the enormous simplification of computation through decimal notation.

Various types of tools have been invented and widely used to assist in numeric calculations. Before Renaissance, they were various types of abaci. More recent examples include slide rules, nomograms and mechanical calculators, such as Pascal's calculator. Presently, they have been supplanted by electronic calculators and computers.

ARITHMETIC OPERATIONS

The basic arithmetic operations are addition, subtraction, multiplication and division, although this subject also includes more advanced operations, such as manipulations of percentages, square roots, exponentiation, and logarithmic functions. Arithmetic is performed according to an order of operations. Any set of objects upon which all four arithmetic operations (except division by 0) can be performed, and where these four operations obey the usual laws, is called a field.

Addition (+)

Addition is the basic operation of arithmetic. In its simplest form, addition combines two numbers, the *addends* or *terms*, into a single number, the *sum* of the numbers (Such as $2 + 2 = 4$ or $3 + 5 = 8$).

Adding more than two numbers can be viewed as repeated addition; this procedure is known as summation and includes ways to add infinitely many numbers in an infinite series; repeated addition of the number 1 is the most basic form of counting.

Addition is commutative and associative so the order the terms are added in does not matter. The identity element of addition (the additive identity) is 0, that is, adding 0 to any number yields that same number. Also, the inverse element of addition (the additive inverse) is the opposite of any number, that is, adding the opposite of any number to the number itself yields the additive identity, 0. For example, the opposite of 7 is −7, so $7 + (-7) = 0$.

Addition can be given geometrically as in the following example:

If we have two sticks of lengths *2* and *5*, then if we place the sticks one after the other, the length of the stick thus formed is $2 + 5 = 7$.

Subtraction (−)

Subtraction is the inverse of addition. Subtraction finds the *difference* between two numbers, the *minuend* minus the *subtrahend*. If the minuend is larger than the subtrahend, the difference is positive; if the minuend is smaller than the subtrahend, the difference is negative; if they are equal, the difference is 0.

Subtraction is neither commutative nor associative. For that reason, it is often helpful to look at subtraction as addition of the minuend and the opposite of the subtrahend, that is $a - b = a + (-b)$. When written as a sum, all the properties of addition hold.

There are several methods for calculating results, some of which are particularly advantageous to machine calculation. For example, digital computers employ the method of two's complement. Of great importance is the counting up method by which change is made. Suppose an amount P is given to pay the required amount Q, with P greater than Q. Rather than performing the subtraction $P - Q$ and counting out that amount in change, money is counted out starting at Q and continuing until reaching P. Although the amount counted out must equal the result of the subtraction $P - Q$, the subtraction was never really done and the value of $P - Q$might still be unknown to the change-maker.

Subtraction is a mathematical operation that represents the operation of removing objects from a collection. It is signified by the minus sign (−). For example, in the picture on the right, there are 5 − 2 apples—meaning 5 apples with 2 taken away, which is a total of 3 apples. Therefore, 5 − 2 = 3. Besides counting fruits, subtraction can also represent combining other physical and abstract quantities using different kinds of objects including negative numbers, fractions, irrational numbers, vectors, decimals, functions, and matrices.

Subtraction follows several important patterns. It is anticommutative, meaning that changing the order changes the sign of the answer. It is not associative, meaning that when one subtracts more than two numbers, the order in which subtraction is performed matters. Subtraction of 0 does not change a number. Subtraction also obeys predictable rules concerning related operations such as addition and multiplication. All of these rules can be proven, starting with the subtraction of integers and generalizing up through the real numbers and beyond. General binary operations that continue these patterns are studied in abstract algebra.

Performing subtraction is one of the simplest numerical tasks. Subtraction of very small numbers is accessible to young children. In primary education, students are taught to subtract numbers in the decimal system, starting with single digits and progressively tackling more difficult problems.

"5 − 2 = 3" (verbally, "five minus two equals three")

$$\begin{array}{r} {}^{6}\not{7}28 \\ -\ 51 \\ \hline 677 \end{array}$$

An example problem

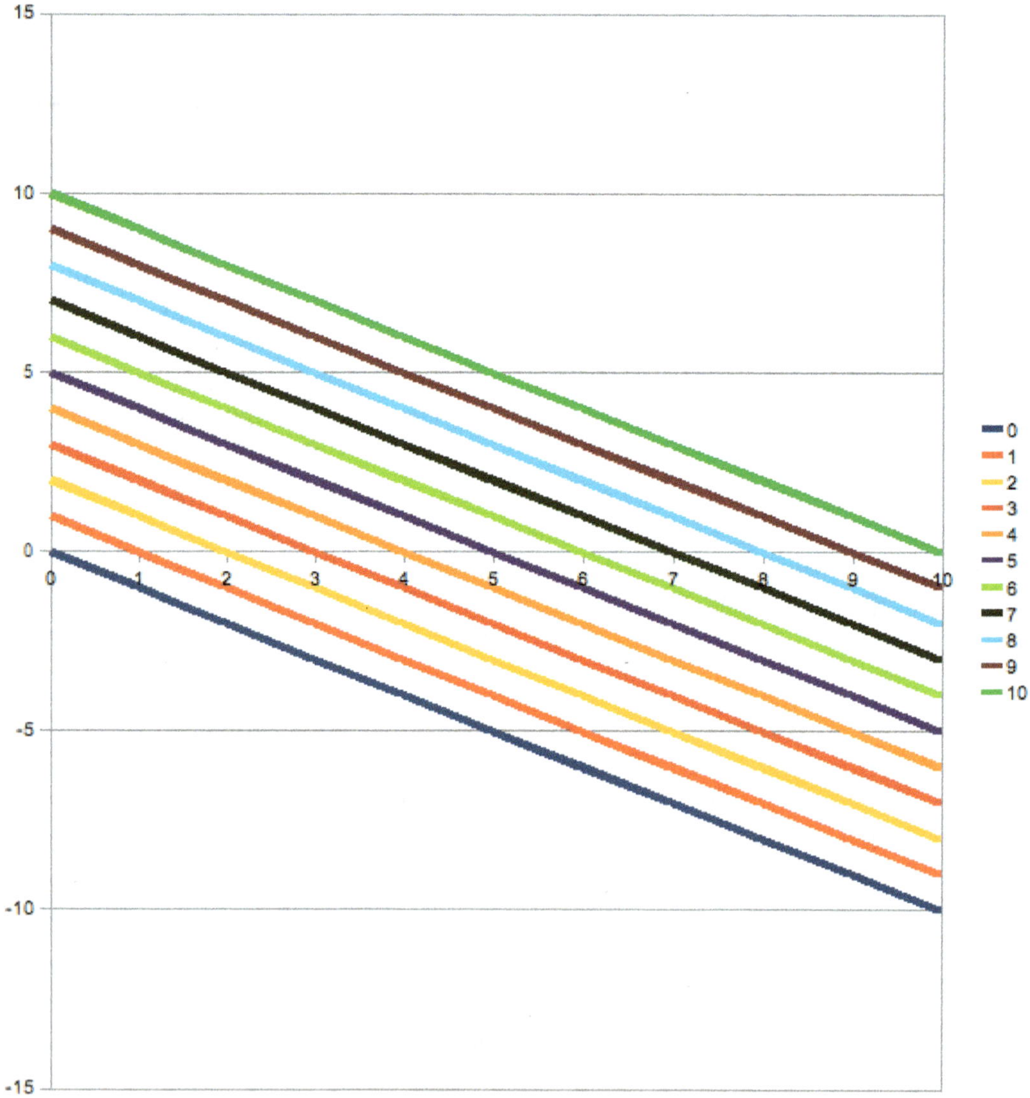

Subtraction of numbers 0–10. Line labels = minuend. X axis = subtrahend. Y axis = difference.

Placard outside shop in Bordeaux advertising subtraction of 20% from the price of a second perfume

Notation and Terminology

Subtraction is written using the minus sign "−" between the terms; that is, in infix notation. The result is expressed with an equals sign. For example,

$2 - 1 = 1$ (verbally, "two minus one equals one")

$4 - 2 = 2$ (verbally, "four minus two equals two")

$6 - 3 = 3$ (verbally, "six minus three equals three")

$4 - 6 = -2$ (verbally, "four minus six equals negative two")

There are also situations where subtraction is "understood" even though no symbol appears:

- A column of two numbers, with the lower number in red, usually indicates that the lower number in the column is to be subtracted, with the difference written below, under a line. This is most common in accounting.

Formally, the number being subtracted is known as the subtrahend, while the number it is subtracted from is the minuend. The result is the difference.

All of this terminology derives from Latin. "Subtraction" is an English word derived from the Latin verb subtrahere, which is in turn a compound of sub "from under" and trahere "to pull"; thus to subtract is to draw from below, take away. Using the gerundive suffix -nd results in "subtrahend", "thing to be subtracted". Likewise from minuere "to reduce or diminish", one gets "minuend", "thing to be diminished".

Integers and Real Numbers

Integers

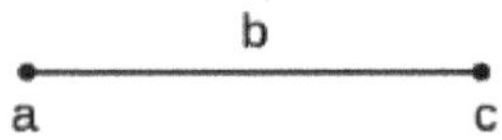

Imagine a line segment of length b with the left end labeled a and the right end labeled c. Starting from a, it takes b steps to the right to reach c. This movement to the right is modeled mathematically by addition:

$a + b = c.$

From c, it takes b steps to the *left* to get back to a. This movement to the left is modeled by subtraction:

$c - b = a.$

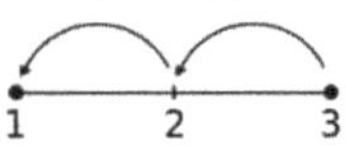

Now, a line segment labeled with the numbers 1, 2, and 3. From position 3, it takes no steps to the left to stay at 3, so $3 - 0 = 3$. It takes 2 steps to the left to get to position 1, so $3 - 2 = 1$. This picture is inadequate to describe what would happen after going 3 steps to the left of position 3. To represent such an operation, the line must be extended.

To subtract arbitrary natural numbers, one begins with a line containing every natural number (0, 1, 2, 3, 4, 5, 6, ...). From 3, it takes 3 steps to the left to get to 0, so $3 - 3 = 0$. But $3 - 4$ is still invalid since it again leaves the line. The natural numbers are not a useful context for subtraction.

The solution is to consider the integer number line (..., −3, −2, −1, 0, 1, 2, 3, ...). From 3, it takes 4 steps to the left to get to −1:

$3 - 4 = -1.$

Natural Numbers

Subtraction of natural numbers is not closed. The difference is not a natural number unless the minuend is greater than or equal to the subtrahend. For example, 26 cannot be subtracted from 11 to give a natural number. Such a case uses one of two approaches:

- Say that 26 cannot be subtracted from 11; subtraction becomes a partial function.
- Give the answer as an integer representing a negative number, so the result of subtracting 26 from 11 is -15.

Real Numbers

Subtraction of real numbers is defined as addition of signed numbers. Specifically, a number is subtracted by adding its additive inverse. Then we have $3 - \pi = 3 + (-\pi)$. This helps to keep the ring of real numbers "simple" by avoiding the introduction of "new" operators such as subtraction. Ordinarily a ring only has two operations defined on it; in the case of the integers, these are addition and multiplication. A ring already has the concept of additive inverses, but it does not have any notion of a separate subtraction operation, so the use of signed addition as subtraction allows us to apply the ring axioms to subtraction without needing to prove anything.

Properties

Anticommutativity

Subtraction is anti-commutative, meaning that if one reverses the terms in a difference left-to-right, the result is the negative of the original result. Symbolically, if a and b are any two numbers, then

a – b = –(b- a)

Non-Associativity

Subtraction is non-associative, which comes up when one tries to define repeated subtraction. Should the expression

“$a - b - c$”

be defined to mean $(a - b) - c$ or $a - (b - c)$? These two possibilities give different answers. To resolve this issue, one must establish an order of operations, with different orders giving different results.

Predecessor

In the context of integers, subtraction of one also plays a special role: for any integer a, the integer $(a - 1)$ is the largest integer less than a, also known as the predecessor of a.

Units of Measurement

When subtracting two numbers with units of measurement such as kilograms or pounds, they must have the same unit. In most cases the difference will have the same unit as the original numbers.

Percentages

Changes in percentages can be reported in at least two forms, percentage change and percentage point change. Percentage change represents the relative change between the two quantities as a percentage, while percentage point change is simply the number obtained by subtracting the two percentages.

As an example, suppose that 30% of widgets made in a factory are defective. Six months later, 20% of widgets are defective. The percentage change is -33 1/3%, while the percentage point change is -10 percentage points.

In Computing

The method of complements is a technique used to subtract one number from another using only addition of positive numbers. This method was commonly used in mechanical calculators and is still used in modern computers. To subtract a binary number *y* (the subtrahend) from another number *x* (the minuend), the ones' complement of *y* is added to *x* and one is added to the sum. The leading digit '1' of the result is then discarded.

The method of complements is especially useful in binary (radix 2) since the ones' complement is very easily obtained by inverting each bit (changing ‹0' to ‹1' and vice versa). And adding 1 to get the two's complement can be done by simulating a carry into the least significant bit. For example:

Binary digit	**Ones' complement**
0	1
1	0

01100100 (x, equals decimal 100)

- 00010110 (y, equals decimal 22)

becomes the sum:

01100100 (x)

+ 11101001 (ones' complement of y)

+ 1 (to get the two's complement)

101001110

Dropping the initial "1" gives the answer: 01001110 (equals decimal 78)

The Teaching of Subtraction in Schools

Methods used to teach subtraction to elementary school vary from country to country, and within a country, different methods are in fashion at different times. In what is, in the United States, called traditional mathematics, a specific process is taught to students at the end of the 1st year or during the 2nd year for use with multi-digit whole numbers, and is extended in either the fourth or fifth grade to include decimal representations of fractional numbers.

In America

Almost all American schools currently teach a method of subtraction using borrowing or regrouping (the decomposition algorithm) and a system of markings called crutches. Although a method of borrowing had been known and published in textbooks previously, the use of crutches in American schools spread after William A. Brownell published a study claiming that crutches were beneficial to students using this method. This system caught on rapidly, displacing the other methods of subtraction in use in America at that time.

In Europe

Some European schools employ a method of subtraction called the Austrian method, also known as the additions method. There is no borrowing in this method. There are also crutches (markings to aid memory), which vary by country.

Comparing the two Main Methods

Both these methods break up the subtraction as a process of one digit subtractions by place value. Starting with a least significant digit, a subtraction of subtrahend:

$s_j\ s_{j-1}\ \dots\ s_1$

from minuend

$m_k\ m_{k-1}\ \dots\ m_1,$

where each s_i and m_i is a digit, proceeds by writing down $m_1 - s_1$, $m_2 - s_2$, and so forth, as long as s_i does not exceed m_i. Otherwise, m_i is increased by 10 and some other digit is modified to correct for this increase. The American method corrects by attempting to decrease the minuend digit m_{i+1} by one (or continuing the borrow leftwards until there is a non-zero digit from which to borrow). The European method corrects by increasing the subtrahend digit s_{i+1} by one.

Example: 704 − 512.

−1			⟵ carry
C	*D*	*U*	
7	0	4	⟵ Minuend
5	1	2	⟵ Subtrahend
1	9	2	⟵ Rest or Difference

The minuend is 704, the subtrahend is 512. The minuend digits are $m_3 = 7$, $m_2 = 0$ and $m_1 = 4$. The subtrahend digits are $s_3 = 5$, $s_2 = 1$ and $s_1 = 2$. Beginning at the one's place, 4 is not less than 2 so the difference 2 is written down in the result's one place. In the ten's place, 0 is less than 1, so the 0 is increased by 10, and the difference with 1, which is 9, is written down in the ten's place. The American method corrects for the increase of ten by reducing the digit in the minuend's hundreds place by one. That is, the 7 is struck through and replaced by a 6. The subtraction then proceeds in the hundreds place, where 6 is not less than 5, so the difference is written down in the result's hundred's place. We are now done, the result is 192.

The Austrian method does not reduce the 7 to 6. Rather it increases the subtrahend hundred's digit by one. A small mark is made near or below this digit (depending on the school). Then the subtraction proceeds by asking what number when increased by 1, and 5 is added to it, makes 7. The answer is 1, and is written down in the result's hundred's place.

There is an additional subtlety in that the student always employs a mental subtraction table in the American method. The Austrian method often encourages the student to mentally use the addition table in reverse. In the example above, rather than adding 1 to 5, getting 6, and subtracting that from 7, the student is asked to consider what number, when increased by 1, and 5 is added to it, makes 7.

SUBTRACTION BY HAND

Austrian Method

Example:

```
 753
-491
----
```

1 + … = 3

```
 753
-491
----
   2
```

The difference is written under the line.

```
 753
-491
----
   2
```

9 + … = 5 The required sum (5) is too small!

```
 753
-491
  1
----
   2
```

So, we add 10 to it and put a 1 under the next higher place in the subtrahend.

```
 753
-491
  1
----
  62
```

9 + … = 15 Now we can find the difference like before.

```
 753
-491
  1
  62
```

(4 + 1) + … = 7

```
 753
-491
 1
 262
```

The difference is written under the line.

```
 753
-491
 1
 262
```

The total difference.

Subtraction from Left to Right

Example:

```
 753
-491
 3
```

7 − 4 = 3 This result is only penciled in.

```
 753
-491
 2
```

Because the next digit of the minuend is smaller than the subtrahend, we subtract one from our penciled-in-number and mentally add ten to the next.

```
 753
-491
 26
```

15 − 9 = 6

```
 753
-491
 26
```

Because the next digit in the minuend is not smaller than the subtrahend, We keep this number.

753
-491
262

3 − 1 = 2

American Method

In this method, each digit of the subtrahend is subtracted from the digit above it starting from right to left. If the top number is too small to subtract the bottom number from it, we add 10 to it; this 10 is 'borrowed' from the top digit to the left, which we subtract 1 from. Then we move on to subtracting the next digit and borrowing as needed, until every digit has been subtracted. Example:

753
-491

3 − 1 = ...

753
-491
2

We write the difference under the line.

753
-491
2

5 − 9 = ...The minuend (5) is too small!

6 15
~~7~~~~5~~3
-491
2

So, we add 10 to it. The 10 is 'borrowed' from the digit on the left, which goes down by 1.

6 15
~~7~~~~5~~3
-491
62

15 − 9 = ...Now the subtraction works, and we write the difference under the line.

6 15
~~7~~~~5~~3
-491
62

6 − 4 = ...

6 15
753
-491
262

We write the difference under the line.

6 15
753
-491
262

The total difference.

Trade First

A variant of the American method where all borrowing is done before all subtraction.

Example:

4 11
751
-493

1 − 3 = not possible.We add a 10 to the 1. Because the 10 is 'borrowed' from the nearby 5, the 5 is lowered by 1.

6 14
4 11
751
-493

4 − 9 = not possible.So we proceed as in step 1.

6 14
4 11
751
-493
8

Working from right to left:11 − 3 = 8

6 14
4 11
751
-493
58

14 − 9 = 5

6 14
4 11
751
-493
258

6 − 4 = 2

Partial Differences

The partial differences method is different from other vertical subtraction methods because no borrowing or carrying takes place. In their place, one places plus or minus signs depending on whether the minuend is greater or smaller than the subtrahend. The sum of the partial differences is the total difference.

Example:

```
 753
-491
----
+300
```

The smaller number is subtracted from the greater:700 − 400 = 300Because the minuend is greater than the subtrahend, this difference has a plus sign.

```
 753
-491
----
+300
-  40
```

The smaller number is subtracted from the greater:90 − 50 = 40 Because the minuend is smaller than the subtrahend, this difference has a minus sign.

```
 753
-491
----
+300
- 40
+  2
```

The smaller number is subtracted from the greater:3 − 1 = 2 Because the minuend is greater than the subtrahend, this difference has a plus sign.

```
 753
-491
----
+300
- 40
+  2
----
 262
```

+ 300 − 40 + 2 = 262

Nonvertical Methods

Counting up

Instead of finding the difference digit by digit, one can count up the numbers between the subtrahend and the minuend.

Example:

1234 − 567 = can be found by the following steps:

- 567 + **3** = 570
- 570 + **30** = 600

- 600 + **400** = 1000
- 1000 + **234** = 1234

Add up the value from each step to get the total difference: 3 + 30 + 400 + 234 = 667.

Breaking Up the Subtraction

Another method that is useful for mental arithmetic is to split up the subtraction into small steps.

Example:

1234 − 567 = can be solved in the following way:

- 1234 − **500** = 734
- 734 − **60** = 674
- 674 − **7** = 667

Same Change

The same change method uses the fact that adding or subtracting the same number from the minuend and subtrahend does not change the answer. One adds the amount needed to get zeros in the subtrahend.

Example:

"1234 − 567 =" can be solved as follows:

- 1234 − 567 = 1237 − 570 = 1267 − 600 = 667

MULTIPLICATION (× OR · OR *)

Multiplication is the second basic operation of arithmetic. Multiplication also combines two numbers into a single number, the *product*. The two original numbers are called the *multiplier* and the *multiplicand*, sometimes both simply called *factors*.

Multiplication may be viewed as a scaling operation. If the numbers are imagined as lying in a line, multiplication by a number, say x, greater than 1 is the same as stretching everything away from 0 uniformly, in such a way that the number 1 itself is stretched to where x was. Similarly, multiplying by a number less than 1 can be imagined as squeezing towards 0. (Again, in such a way that 1 goes to the multiplicand.)

Multiplication is commutative and associative; further it is distributive over addition and subtraction. The multiplicative identity is 1, that is, multiplying any number by 1 yields that same number. Also, the multiplicative inverse is the reciprocal of any number (except 0; 0 is the only number without a multiplicative inverse), that is, multiplying the reciprocal of any number by the number itself yields the multiplicative identity.

The product of a and b is written as $a \times b$ or $a \cdot b$. When a or b are expressions not written simply with digits, it is also written by simple juxtaposition: ab. In computer programming languages and software packages in which one can only use characters normally found on a keyboard, it is often written with an asterisk: $a * b$.

Multiplication (often denoted by the cross symbol “×», by a point “·», by juxtaposition, or, on computers, by an asterisk “*”) is one of the four elementary, mathematical operations of arithmetic; with the others being addition, subtraction and division.

The multiplication of whole numbers may be thought as a repeated addition; that is, the multiplication of two numbers is equivalent to adding as many copies of one of them, the *multiplicand*, as the value of the other one, the *multiplier*. Normally, the multiplier is written first and multiplicand second, though this can vary, as the distinction is not very meaningful:

$$a \times b = \underbrace{b + \cdots + b}_{a} = \underbrace{a + \cdots + a}_{b}$$

For example, 4 multiplied by 3 (often written as 3×4 and said as "3 times 4") can be calculated by adding 3 copies of 4 together:

$$3 \times 4 = 4 + 4 + 4 = 12$$

Here 3 and 4 are the “factors” and 12 is the “product”.

One of the main properties of multiplication is the commutative property, adding 3 copies of 4 gives the same result as adding 4 copies of 3:

$$4 \times 3 = 3 + 3 + 3 + 3 = 12$$

The multiplication of integers (including negative numbers), rational numbers (fractions) and real numbers is defined by a systematic generalization of this basic definition.

Multiplication can also be visualized as counting objects arranged in a rectangle (for whole numbers) or as finding the area of a rectangle whose sides have given lengths. The area of a rectangle does not depend on which side is measured first, which illustrates the commutative property. The product of two measurements is a new type of measurement, for instance multiplying the lengths of the two sides of a rectangle gives its area, this is the subject of dimensional analysis.

The inverse operation of multiplication is division. For example, since 4 multiplied by 3 equals 12, then 12 divided by 3 equals 4. Multiplication by 3, followed by division by 3, yields the original number (since the division of a number other than 0 by itself equals 1).

Multiplication is also defined for other types of numbers, such as complex numbers, and more abstract constructs, like matrices. For these more abstract constructs, the order that the operands are multiplied sometimes does matter. A listing of the many different kinds of products that are used in mathematics is given in the product (mathematics) page.

Four bags with three marbles per bag gives twelve marbles (4 × 3 = 12).

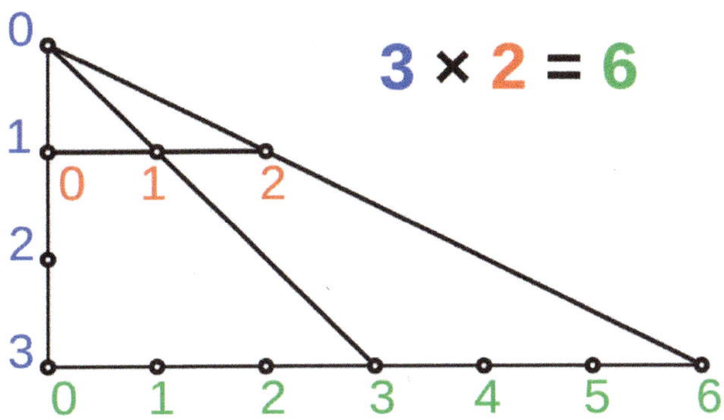

Multiplication can also be thought of as scaling. Here we see 2 being multiplied by 3 using scaling, giving 6 as a result

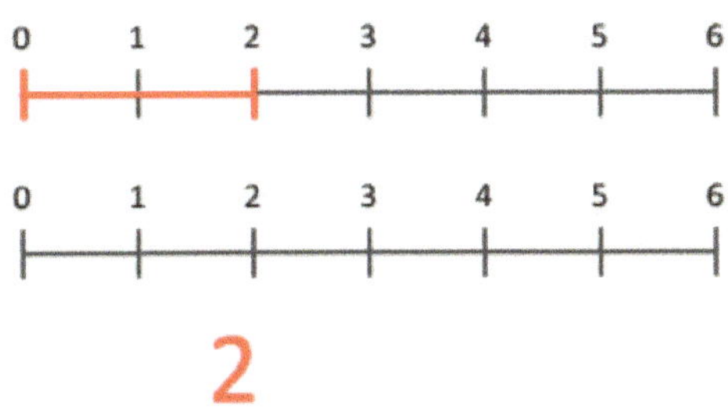

Animation for the multiplication 2 × 3 = 6.

4 × 5 = 20, the rectangle is composed of 20 squares, each having dimensions of 1 by 1.

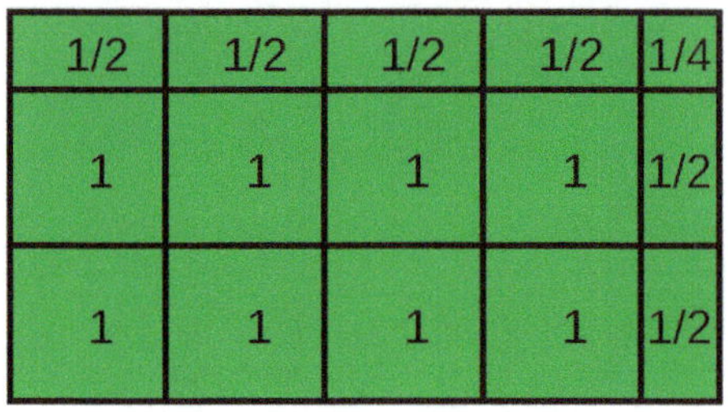

Area of a cloth 4.5m × 2.5m = 11.25m^2; 4½ × 2½ = 11¼

Notation and Terminology

In arithmetic, multiplication is often written using the sign "×" between the terms; that is, in infix notation. For example,

$2 \times 3 = 6$ (verbally, "two times three equals six")

$3 \times 4 = 12$

$2 \times 3 \times 5 = 6 \times 5 = 30$

$2 \times 2 \times 2 \times 2 \times 2 = 32$

The multiplication sign ×

The sign is encoded in Unicode at U+00D7 × MULTIPLICATION SIGN (HTML × ·×).

There are other mathematical notations for multiplication:

- Multiplication is also denoted by dot signs, usually a middle-position dot (rarely period):

$$5 \cdot 2 \quad \text{or} \quad 5 . 2$$

The middle dot notation, encoded in Unicode as U+22C5 · DOT OPERATOR, is standard in the United States, the United Kingdom, and other countries where the period is used as a decimal point. When the dot operator character is not accessible, the interpunct (·) is used. In other countries that use a comma as a decimal mark, either the period or a middle dot is used for multiplication.

- In algebra, multiplication involving variables is often written as a juxtaposition (e.g., xy for x times y or $5x$ for five times x). The notation can also be used for quantities that are surrounded by parentheses (e.g., 5(2) or (5)(2) for five times two). This implicit usage of multiplication can cause ambiguity when the concatenated variables happen to match the name of another variable, when a variable name in front of a parenthesis can be confused with a function name, or in the correct determination of the order of operations.
- In matrix multiplication, there is a distinction between the cross and the dot symbols. The cross symbol generally denotes the taking a cross product of two vectors, yielding a vector as the result, while the dot denotes taking the dot product of two vectors, resulting in a scalar.

In computer programming, the asterisk (as in 5*2) is still the most common notation. This is due to the fact that most computers historically were limited to small character sets (such as ASCII and EBCDIC) that lacked a multiplication sign (such as • or ×), while the asterisk appeared on every keyboard. This usage originated in the FORTRAN programming language. The numbers to be multiplied are generally called the "factors". The number to be multiplied is called the "multiplicand", while the number of times the multiplicand is to be multiplied comes from the "multiplier". Usually the multiplier is placed first and the multiplicand is placed second, however sometimes the first factor is the multiplicand and the second the multiplier.Additionally, there are some sources in which the term "multiplicand" is regarded as a synonym for "factor". In algebra, a number that is the multiplier of a variable or expression (e.g., the 3 in $3xy^2$) is called a coefficient.

The result of a multiplication is called a product. A product of integers is a multiple of each factor. For example, 15 is the product of 3 and 5, and is both a multiple of 3 and a multiple of 5.

Computation

The common methods for multiplying numbers using pencil and paper require a multiplication table of memorized or consulted products of small numbers (typically any two numbers from 0 to 9), however one method, the peasant multiplication algorithm, does not.

Multiplying numbers to more than a couple of decimal places by hand is tedious and error prone. Common logarithms were invented to simplify such calculations. The slide rule allowed numbers to be quickly multiplied to about three places of accuracy.

Beginning in the early twentieth century, mechanical calculators, such as the Marchant, automated multiplication of up to 10 digit numbers. Modern electronic computers and calculators have greatly reduced the need for multiplication by hand.

Historical Algorithms

Methods of multiplication were documented in the Egyptian, Greek, Indian and Chinese civilizations.

The Ishango bone, dated to about 18,000 to 20,000 BC, hints at a knowledge of multiplication in thc Uppcr Palcolithic cra in Ccntral Africa.

Egyptians

The Egyptian method of multiplication of integers and fractions, documented in the Ahmes Papyrus, was by successive additions and doubling. For instance, to find the product of 13 and 21 one had to double 21 three times, obtaining $2 \times 21 = 42$, $4 \times 21 = 2 \times 42 = 84$, $8 \times 21 = 2 \times 84 = 168$. The full product could then be found by adding the appropriate terms found in the doubling sequence:

$13 \times 21 = (1 + 4 + 8) \times 21 = (1 \times 21) + (4 \times 21) + (8 \times 21) = 21 + 84 + 168 = 273.$

Babylonians

The Babylonians used a sexagesimal positional number system, analogous to the modern day decimal system. Thus, Babylonian multiplication was very similar to modern decimal multiplication. Because of the relative difficulty of remembering 60×60 different products, Babylonian mathematicians employed multiplication tables. These tables consisted of a list of the first twenty multiples of a certain *principal number n*: n, $2n$, ..., $20n$; followed by the multiples of $10n$: $30n$ $40n$, and $50n$. Then to compute any sexagesimal product, say $53n$, one only needed to add $50n$ and $3n$ computed from the table.

Chinese

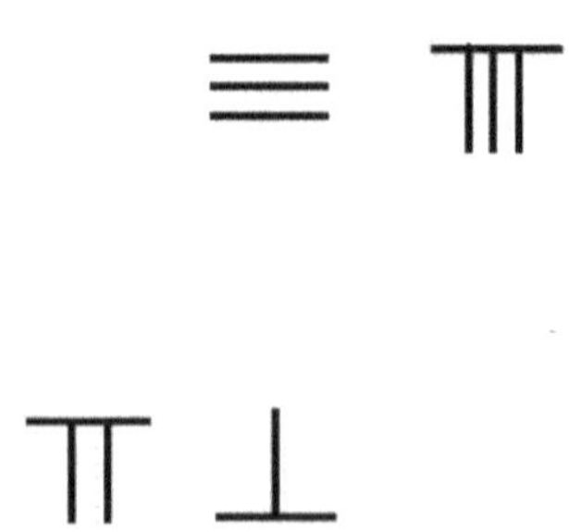

$38 \times 76 = 2888$

In the mathematical text *Zhou Bi Suan Jing*, dated prior to 300 BC, and the *Nine Chapters on the Mathematical Art*, multiplication calculations were written out in words, although the early Chinese mathematicians employed Rod calculus involving place value addition, subtraction, multiplication and division. These place value decimal arithmetic algorithms were introduced by Al Khwarizmi to Arab countries in the early 9th century.

Modern Methods

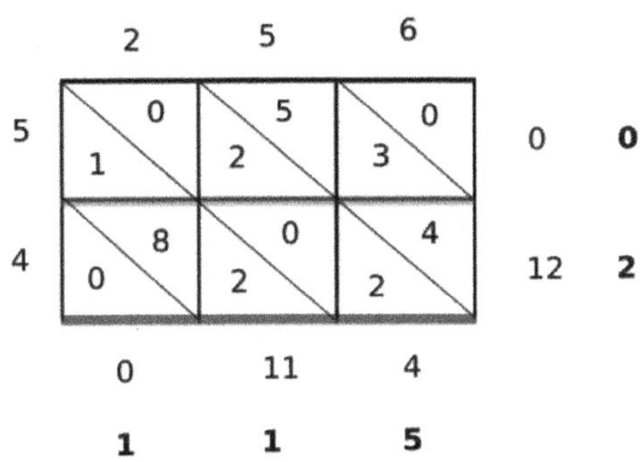

Product of 45 and 256. Note the order of the numerals in 45 is reversed down the left column. The carry step of the multiplication can be performed at the final stage of the calculation (in bold), returning the final product of 45 × 256 = 11520. This is a variant of Lattice multiplication.

The modern method of multiplication based on the Hindu–Arabic numeral system was first described by Brahmagupta. Brahmagupta gave rules for addition, subtraction, multiplication and division. Henry Burchard Fine, then professor of Mathematics at Princeton University, wrote the following:

The Indians are the inventors not only of the positional decimal system itself, but of most of the processes involved in elementary reckoning with the system. Addition and subtraction they performed quite as they are performed nowadays; multiplication they effected in many ways, ours among them, but division they did cumbrously.

Grid Method

Grid method multiplication or the box method, is used in primary schools in England and Wales to help teach and understanding of how multiple digit multiplication works. An example of multiplying 34 by 13 would be to lay the numbers out in a grid like:

	30	4
10	300	40
3	90	12

and then add the entries.

Computer Algorithms

The classical method of multiplying two n-digit numbers requires n^2 simple multiplications. Multiplication algorithms have been designed that reduce the computation time considerably when multiplying large numbers. In particular for very large numbers methods based on the Discrete Fourier Transform can reduce the number of simple multiplications to the order of $n \log_2(n) \log_2 \log_2(n)$.

Products of Measurements

One can only meaningfully add or subtract quantities of the same type but can multiply or divide quantities of different types. Four bags with three marbles each can be though of as:

[4 bags] × [3 marbles per bag] = 12 marbles.

When two measurements are multiplied together the product is of a type depending on the types of the measurements. The general theory is given by dimensional analysis. This analysis is routinely applied in physics but has also found applications in finance.

A common example is multiplying speed by time gives distance, so

50 kilometers per hour × 3 hours = 150 kilometers.

Other examples:

$$2.5 \text{ meters} \times 4.5 \text{ meters} = 11.25 \text{ square meters}$$
$$11 \text{ meters/second} \times 9 \text{ seconds} = 99 \text{ meters}$$

Products of Sequences

Capital Pi Notation

The product of a sequence of terms can be written with the product symbol, which derives from the capital letter Π (Pi) in the Greek alphabet. Unicode position U+220F (∏) contains a glyph for denoting such a product, distinct from U+03A0 (Π), the letter. The meaning of this notation is given by:

$$\prod_{i=1}^{4} i = 1 \cdot 2 \cdot 3 \cdot 4,$$

that is

$$\prod_{i=1}^{4} i = 24.$$

The subscript gives the symbol for a dummy variable (*i* in this case), called the "index of multiplication" together with its lower bound (*1*), whereas the superscript (here *4*) gives its upper bound. The lower and upper bound are expressions denoting integers. The factors of the product are obtained by taking the expression following the product operator, with successive integer values substituted for the index of multiplication, starting from the lower bound and incremented by 1 up to and including the upper bound. So, for example:

$$\prod_{i=1}^{6} i = 1 \cdot 2 \cdot 3 \cdot 4 \cdot 5 \cdot 6 = 720$$

More generally, the notation is defined as

$$\prod_{i=m}^{n} x_i = x_m \cdot x_{m+1} \cdot x_{m+2} \cdot \quad \cdots \quad \cdot x_{n-1} \cdot x_n,$$

where *m* and *n* are integers or expressions that evaluate to integers. In case $m = n$, the value of the product is the same as that of the single factor x_m. If $m > n$, the product is the empty product, with the value 1.

Infinite Products

One may also consider products of infinitely many terms; these are called infinite products. Notationally, we would replace *n* above by the lemniscate ∞. The product of such a series is defined as the limit of the product of the first *n* terms, as *n* grows without bound. That is, by definition,

$$\prod_{i=m}^{\infty} x_i = \lim_{n\to\infty} \prod_{i=m}^{n} x_i.$$

One can similarly replace *m* with negative infinity, and define:

$$\prod_{i=-\infty}^{\infty} x_i = \left(\lim_{m\to-\infty} \prod_{i=m}^{0} x_i\right) \cdot \left(\lim_{n\to\infty} \prod_{i=1}^{n} x_i\right),$$

provided both limits exist.

Properties

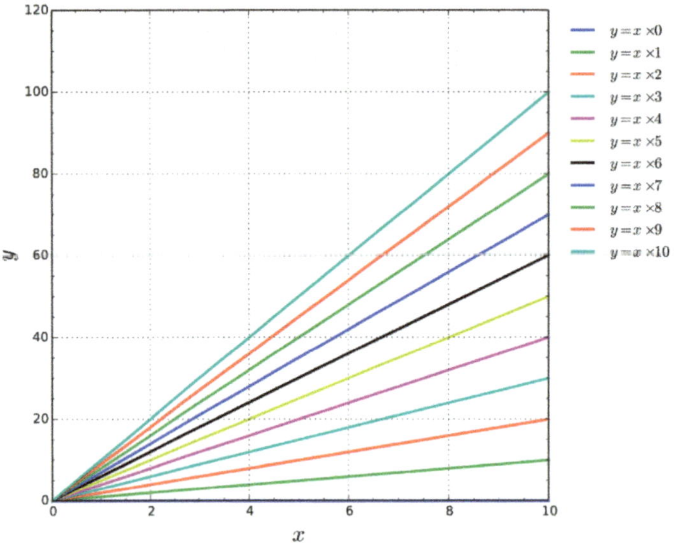

Multiplication of numbers 0–10. Line labels = multiplicand. X axis = multiplier. Y axis = product. Extension of this pattern into other quadrants gives the reason why a negative number times a negative number yields a positive number. Note also how multiplication by zero causes a reduction in dimensionality, as does multiplication by a singular matrix where the determinant is 0. In this process, information is lost and cannot be regained.

For the real and complex numbers, which includes for example natural numbers, integers and fractions, multiplication has certain properties:

Commutative property

The order in which two numbers are multiplied does not matter:

$$x \cdot y = y \cdot x.$$

Associative property

Expressions solely involving multiplication or addition are invariant with respect to order of operations:

$$(x \cdot y) \cdot z = x \cdot (y \cdot z)$$

Distributive property

Holds with respect to multiplication over addition. This identity is of prime importance in simplifying algebraic expressions:

$$x \cdot (y + z) = x \cdot y + x \cdot z$$

Identity element

The multiplicative identity is 1; anything multiplied by one is itself. This is known as the identity property:

$$x \cdot 1 = x$$

Negation

Negative one times any number is equal to the additive inverse of that number.

$$(-1) \cdot x = (-x)$$

Negative one times negative one is positive one.

$$(-1) \cdot (-1) = 1$$

The natural numbers do not include negative numbers.

Inverse element

Every number *x*, except zero, has a multiplicative inverse $\frac{1}{x}$, such that $x \cdot \left(\frac{1}{x}\right) = 1.$

Order preservation

Multiplication by a positive number preserves order:

if $a > 0$, then if $b > c$ then $ab > ac$.

Multiplication by a negative number reverses order:

if $a < 0$ and $b > c$ then $ab < ac$.

The complex numbers do not have an order predicate.

Other mathematical systems that include a multiplication operation may not have all these properties. For example, multiplication is not, in general, commutative for matrices and quaternions.

Axioms

In the book *Arithmetices principia, nova methodo exposita*, Giuseppe Peano proposed axioms for arithmetic based on his axioms for natural numbers. Peano arithmetic has two axioms for multiplication:

$$x \times 0 = 0$$

$$x \times S(y) = (x \times y) + x$$

Here *S*(*y*) represents the successor of *y*, or the natural number that *follows y*. The various properties like associativity can be proved from these and the other axioms of Peano arithmetic including induction. For instance *S*(0). denoted by 1, is a multiplicative identity because

$$x \times 1 = x \times S(0) = (x \times 0) + x = 0 + x = x$$

The axioms for integers typically define them as equivalence classes of ordered pairs of natural numbers. The model is based on treating (*x*,*y*) as equivalent to *x*−*y* when *x* and *y* are treated as integers. Thus both (0,1) and (1,2) are equivalent to −1. The multiplication axiom for integers defined this way is

$$(x_p, x_m) \times (y_p, y_m) = (x_p \times y_p + x_m \times y_m,\ x_p \times y_m + x_m \times y_p)$$

The rule that −1 × −1 = 1 can then be deduced from

$$(0,1) \times (0,1) = (0 \times 0 + 1 \times 1,\ 0 \times 1 + 1 \times 0) = (1,0)$$

Multiplication is extended in a similar way to rational numbers and then to real numbers.

Multiplication with Set Theory

The product of non-negative integers can be defined with set theory using cardinal numbers or the Peano axioms. See below how to extend this to multiplying arbitrary integers, and then arbitrary rational numbers. The product of real numbers is defined in terms of products of rational numbers, see construction of the real numbers.

Multiplication in Group Theory

There are many sets that, under the operation of multiplication, satisfy the axioms that define group structure. These axioms are closure, associativity, and the inclusion of an identity element and inverses.

A simple example is the set of non-zero rational numbers. Here we have identity 1, as opposed to groups under addition where the identity is typically 0. Note that with the rationals, we must exclude zero because, under multiplication, it does not have an inverse: there is no rational number that can be multiplied by zero to result in 1. In this example we have an abelian group, but that is not always the case.

To see this, look at the set of invertible square matrices of a given dimension, over a given field. Now it is straightforward to verify closure, associativity, and inclusion of identity (the identity matrix) and inverses. However, matrix multiplication is not commutative, therefore this group is nonabelian. Another fact of note is that the integers under multiplication is not a group, even if we exclude zero. This is easily seen by the nonexistence of an inverse for all elements other than 1 and −1.

Multiplication in group theory is typically notated either by a dot, or by juxtaposition (the omission of an operation symbol between elements). So multiplying element **a** by element **b** could be notated **a** .**b** or **ab**. When referring to a group via the indication of the set and operation, the dot is used, e.g., our first example could be indicated by $(\mathbb{Q} \setminus \{0\}, \cdot)$

Multiplication of Different Kinds of Numbers

Numbers can *count* (3 apples), *order* (the 3rd apple), or *measure* (3.5 feet high); as the history of mathematics has progressed from counting on our fingers to modelling quantum mechanics, multiplication has been generalized to more complicated and abstract types of numbers, and to things that are not numbers (such as matrices) or do not look much like numbers (such as quaternions).

Integers

$N \times M$ is the sum of *M* copies of *N* when *N* and *M* are positive whole numbers. This gives the number of things in an array *N* wide and *M* high. Generalization to negative numbers can be done by
$N \times (-M) = (-N) \times M = -(N \times M)$ and
$(-N) \times (-M) = N \times M$
The same sign rules apply to rational and real numbers.

Rational numbers

Generalization to fractions $\frac{A}{B} \times \frac{C}{D}$ is by multiplying the numerators and denominators respectively:
$\frac{A}{B} \times \frac{C}{D} = \frac{(A \times C)}{(B \times D)}$. This gives the area of a rectangle $\frac{A}{B}$ high and $\frac{C}{D}$ wide, and is the same as the number of things in an array when the rational numbers happen to be whole numbers.

Real numbers

Real numbers and their products can be defined in terms of sequences of rational numbers.

Complex numbers

Considering complex numbers z_1 and z_2 as ordered pairs of real numbers (a_1, b_1) and (a_2, b_2), the product $z_1 \times z_2$ is $(a_1 \times a_2 - b_1 \times b_2, a_1 \times b_2 + a_2 \times b_1)$. This is the same as for reals, $a_1 \times a_2$, when the *imaginary parts* b_1 and b_2 are zero.

Equivalently, denoting $\sqrt{-1}$ as i, we have
$z_1 \times z_2 = (a_1 + b_1 i)(a_2 + b_2 i) = (a_1 \times a_2) + (a_1 \times b_2 i) + (b_1 \times a_2 i) + (b_1 \times b_2 i^2) = (a_1 a_2 - b_1 b_2) + (a_1 b_2 + b_1 a_2)i.$

Further generalizations

See Multiplication in group theory, above, and Multiplicative Group, which for example includes matrix multiplication. A very general, and abstract, concept of multiplication is as the "multiplicatively denoted" (second) binary operation in a ring. An example of a ring that is not any of the above number systems is a polynomial ring (you can add and multiply polynomials, but polynomials are not numbers in any usual sense.)

Division

Often division, $\frac{x}{y}$, is the same as multiplication by an inverse, $x\left(\frac{1}{y}\right)$. Multiplication for some types of "numbers" may have corresponding division, without inverses; in an integral domain x may have no inverse "$\frac{1}{x}$" but $\frac{x}{y}$ may be defined. In a division ring there are inverses, but $\frac{x}{y}$ may be ambiguous in non-commutative rings since $x\left(\frac{1}{y}\right)$ need not the same as $\left(\frac{1}{y}\right)x$

Exponentiation

When multiplication is repeated, the resulting operation is known as **exponentiation**. For instance, the product of three factors of two (2×2×2) is "two raised to the third power", and is denoted by 2^3, a two with a superscript three. In this example, the number two is the **base**, and three is the **exponent**. In general, the exponent (or superscript) indicates how many times the base appears in the expression, so that the expression

$$a^n = \underbrace{a \times a \times \cdots \times a}_{n}$$

indicates that *n* copies of the base *a* are to be multiplied together. This notation can be used whenever multiplication is known to be power associative.

Division (÷ Or /)

Division is essentially the inverse of multiplication. Division finds the *quotient* of two numbers, the *dividend* divided by the *divisor*. Any dividend divided by 0 is undefined. For distinct positive numbers, if the dividend is larger than the divisor, the quotient is greater than 1, otherwise it is less than 1 (a similar rule applies for negative numbers). The quotient multiplied by the divisor always yields the dividend.

Division is neither commutative nor associative. As it is helpful to look at subtraction as addition, it is helpful to look at division as multiplication of the dividend times the reciprocal of the divisor, that is $a \div b = a \times {}^{1}\!/_{b}$. When written as a product, it obeys all the properties of multiplication.

DIVISION (MATHEMATICS)

Division is one of the four basic operations of arithmetic, the others being addition, subtraction, and multiplication. The division of two natural numbers is the process of calculating the number of times one number is contained within one another.[7] For example, in the picture on the right, the 20 apples are divided into groups of five apples, and there exist four groups, meaning that five can be contained within 20 four times, or 20 ÷ 5 = 4. Division can also be thought of as the process of evaluating a fraction, and fractional notation (a/b and $^a/_b$) is commonly used to represent division.

Division is the inverse of multiplication; if $a \times b = c$, then $a = c \div b$, as long as b is not zero. Division by zero is undefined for the real numbers and most other contexts,[:246] because if $b = 0$, then a cannot be deduced from b and c, as then c will always equal zero regardless of a. In some contexts, division by zero can be defined although to a limited extent, and limits involving division of a real number as it approaches zero are defined.[a]

In division, the *dividend* is divided by the *divisor* to get a *quotient.* In the above example, 20 is the dividend, five is the divisor, and the quotient is four. In some cases, the divisor may not be contained fully by the dividend; for example, 10 ÷ 3 leaves a remainder of 1/3 as 10 is not a multiple of three. Normally, this remainder is added to the quotient so 10 ÷ 3 would equal 31/3 or 3.33 ..., but in the context of integer division, where numbers have no fractional part, the remainder is discarded.

Besides dividing apples, division can be applied to other physical and abstract objects. Division has been defined in several contexts, such as for the real and complex numbers and for more abstract contexts such as for vector spaces and fields.

Teaching division usually leads to the concept of fractions being introduced to school pupils. Unlike addition, subtraction, and multiplication, the set of all integers is not closed under division. Dividing two integers may result in a remainder. To complete the division of the remainder, the number system is extended to include fractions or rational numbers as they are more generally called.

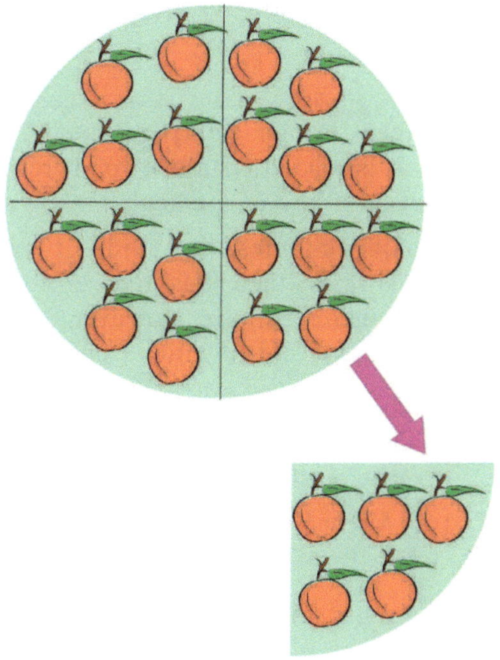

20 ÷ 5 = 4, because 20 apples contain five apples four times.

Notation

Division is often shown in algebra and science by placing the *dividend* over the *divisor* with a horizontal line, also called a fraction bar, between them. For example, *a* divided by *b* is written

$$\frac{a}{b}$$

This can be read out loud as "*a* divided by *b*", "*a* by *b*" or "*a* over *b*". A way to express division all on one line is to write the *dividend* (or numerator), then a slash, then the *divisor* (or denominator), like this:

a/b

This is the usual way to specify division in most computer programming languages since it can easily be typed as a simple sequence of ASCII characters. Some mathematical software, such as MATLAB and GNU Octave, allows the operands to be written in the reverse order by using the backslash as the division operator:

b\a

A typographical variation halfway between these two forms uses a solidus (fraction slash) but elevates the dividend, and lowers the divisor:

$^{a}/_{b}$

Any of these forms can be used to display a fraction. A fraction is a division expression where both dividend and divisor are integers (typically called the *numerator* and *denominator*), and there is no implication that the division must be evaluated further. A second way to show division is to use the obelus (or division sign), common in arithmetic, in this manner:

a/b

This form is infrequent except in elementary arithmetic. ISO 80000-2-9.6 states it should not be used. The obelus is also used alone to represent the division operation itself, as for instance as a label on a key of a calculator.

In some non-English-speaking cultures, "a divided by b" is written *a* : *b*. This notation was introduced in 1631 by William Oughtred in his *Clavis Mathematicae* and later popularized by Gottfried Wilhelm Leibniz. However, in English usage the colon is restricted to expressing the related concept of ratios (then "*a* is to *b*").

In elementary classes of some countries, the notation $b)\,a$ or $b\overline{)a}$ is used to denote *a* divided by *b*, especially when discussing long division; similarly, commonly used in Latin America, $\underline{b)a}$ for short division (as shown in an example on that page). This notation was first introduced by Michael Stifel in *Arithmetica integra*, published in 1544.

Computing

Manual Methods

Division is often introduced through the notion of "sharing out" a set of objects, for example a pile of sweets, into a number of equal portions. Distributing the objects several at a time in each round of sharing to each portion leads to the idea of "chunking", i.e., division by repeated subtraction.

More systematic and more efficient (but also more formalised and more rule-based, and more removed from an overall holistic picture of what division is achieving), a person who knows the multiplication tables can divide two integers using pencil and paper using the method of short division, if the divisor is simple. Long division is used for larger integer divisors. If the dividend has a fractional part (expressed as a decimal fraction), one can continue the algorithm past the ones place as far as desired. If the divisor has a fractional part, we can restate the problem by moving the decimal to the right in both numbers until the divisor has no fraction.

A person can calculate division with an abacus by repeatedly placing the dividend on the abacus, and then subtracting the divisor the offset of each digit in the result, counting the number of divisions possible at each offset.

A person can use logarithm tables to divide two numbers, by subtracting the two numbers' logarithms, then looking up the antilogarithm of the result.

A person can calculate division with a slide rule by aligning the divisor on the C scale with the dividend on the D scale. The quotient can be found on the D scale where it is aligned with the left index on the C scale. The user is responsible, however, for mentally keeping track of the decimal point.

By Computer or with Computer Assistance

Modern computers compute division by methods that are faster than long division: see Division algorithm.

In modular arithmetic, some numbers have a multiplicative inverse with respect to the modulus. We can calculate division by multiplication in such a case. This approach is useful in computers that do not have a fast division instruction.

Properties

Division is right-distributive over addition and subtraction. That means:

$$\frac{a+b}{c} = (a+b) \div c = \frac{a}{c} + \frac{b}{c}$$

in the same way as in multiplication $(a+b) \times c = a \times c + b \times c$ But division is not left-distributive, i.e. we have $\frac{a}{b+c} = a \div (b+c) \neq \frac{a}{b} + \frac{a}{c}$ unlike multiplication.

Euclidean Division

The Euclidean division is the mathematical formulation of the outcome of the usual process of division of integers. It asserts that, given two integers, a, the *dividend*, and b, the *divisor*, such that $b \neq 0$, there are unique integers q, the *quotient*, and r, the remainder, such that $a = bq + r$ and $0 \leq r < | b |$, where $| b |$ denotes the absolute value of b.

Of Integers

Division of integers is not closed. Apart from division by zero being undefined, the quotient is not an integer unless the dividend is an integer multiple of the divisor. For example, 26 cannot be divided by 11 to give an integer. Such a case uses one of five approaches:

- Say that 26 cannot be divided by 11; division becomes a partial function.
- Give an approximate answer as a decimal fraction or a mixed number, so $\frac{26}{11} \simeq 2.36$ or $\frac{26}{11} \simeq 2\frac{36}{100}$. This is the approach usually taken in numerical computation.
- Give the answer as a fraction representing a rational number, so the result of the division of 26 by 11 is $\frac{26}{11}$. But, usually, the resulting fraction should be simplified: the result of the division of 52 by 22 is also $\frac{26}{11}$. This simplification may be done by factoring out the greatest common divisor.
- Give the answer as an integer *quotient* and a *remainder*, so $\frac{26}{11} = 2$ remainder 4. o make the distinction with the previous case, this division, with two integers as result, is sometimes called *Euclidean division*, because it is the basis of the Euclidean algorithm.
- Give the integer quotient as the answer, so $\frac{26}{11} = 2$. This is sometimes called *integer division*.

Dividing integers in a computer program requires special care. Some programming languages, such as C, treat integer division as in case 5 above, so the answer is an integer. Other languages, such as MATLAB and every computer algebra system return a rational number as the answer, as in case 3 above. These languages also provide functions to get the results of the other cases, either directly or from the result of case 3.

Names and symbols used for integer division include div, /, \, and %. Definitions vary regarding integer division when the dividend or the divisor is negative: rounding may be toward zero (so called T-division) or toward $-\infty$ (F-division); rarer styles can occur – see Modulo operation for the details.

Divisibility rules can sometimes be used to quickly determine whether one integer divides exactly into another.

Of Rational Numbers

The result of dividing two rational numbers is another rational number when the divisor is not 0. The division of two rational numbers p/q and r/s can be computed as

$$\frac{p/q}{r/s} = \frac{p}{q} \times \frac{s}{r} = \frac{ps}{qr}.$$

All four quantities are integers, and only p may be 0. This definition ensures that division is the inverse operation of multiplication.

Of Real Numbers

Division of two real numbers results in another real number when the divisor is not 0. It is defined such $a/b = c$ if and only if $a = cb$ and $b \neq 0$.

By Zero

Division of any number by zero (where the divisor is zero) is undefined. This is because zero multiplied by any finite number always results in a product of zero. Entry of such an expression into most calculators produces an error message.

Of Complex Numbers

Dividing two complex numbers results in another complex number when the divisor is not 0, which is found using the conjugate of the denominator:

$$\frac{p+iq}{r+is} = \frac{(p+iq)(r-is)}{(r+is)(r-is)} = \frac{pr+qs+i(qr-ps)}{r^2+s^2} = \frac{pr+qs}{r^2+s^2} + i\frac{qr-ps}{r^2+s^2}.$$

This process of multiplying and dividing by $r - is$ is called 'realisation' or (by analogy) rationalisation. All four quantities p, q, r, s are real numbers, and r and s may not both be 0.

Division for complex numbers expressed in polar form is simpler than the definition above:

$$\frac{pe^{iq}}{re^{is}} = \frac{pe^{iq}e^{-is}}{re^{is}e^{-is}} = \frac{p}{r}e^{i(q-s)}.$$

Again all four quantities p, q, r, s are real numbers, and r may not be 0.

Of Polynomials

One can define the division operation for polynomials in one variable over a field. Then, as in the case of integers, one has a remainder. See Euclidean division of polynomials, and, for hand-written computation, polynomial long division or synthetic division.

Of Matrices

One can define a division operation for matrices. The usual way to do this is to define $A / B = AB^{-1}$, where B^{-1} denotes the inverse of B, but it is far more common to write out AB^{-1} explicitly to avoid confusion. An elementwise division can also be defined in terms of the Hadamard product.

Left and Right Division

Because matrix multiplication is not commutative, one can also define a left division or so-called *backslash-division* as $A \setminus B = A^{-1}B$. For this to be well defined, B^{-1} need not exist, however A^{-1} does need to exist. To avoid confusion, division as defined by $A / B = AB^{-1}$ is sometimes called *right division* or *slash-division* in this context.

Note that with left and right division defined this way, $A/(BC)$ is in general not the same as $(A/B)/C$ and nor is $(AB)\setminus C$ the same as $A\setminus(B\setminus C)$, but $A/(BC) = (A/C)/B$ and $(AB)\setminus C = B\setminus(A\setminus C)$.

Pseudoinverse

To avoid problems when A^{-1} and/or B^{-1} do not exist, division can also be defined as multiplication with the pseudoinverse, i.e., $A / B = AB^{+}$ and $A \setminus B = A^{+}B$, where A^{+} and B^{+} denote the pseudoinverse of A and B.

Decimal Arithmetic

Decimal representation refers exclusively, in common use, to the written numeral system employing arabic numerals as the digits for a radix 10 ("decimal")positional notation; however, any numeral system based on powers of 10, e.g., Greek, Cyrillic, Roman, or Chinese numerals may conceptually be described as "decimal notation" or "decimal representation".

Modern methods for four fundamental operations (addition, subtraction, multiplication and division) were first devised by Brahmagupta of India. This was known during medieval Europe as "Modus Indoram" or Method of the Indians. Positional notation (also known as "place-value notation") refers to the representation or encoding of numbers using the same symbol for the different orders of magnitude (e.g., the "ones place", "tens place", "hundreds place") and, with a radix point, using those same symbols to represent fractions (e.g., the "tenths place", "hundredths place"). For example, 507.36 denotes 5 hundreds (10^2), plus 0 tens (10^1), plus 7 units (10^0), plus 3 tenths (10^{-1}) plus 6 hundredths (10^{-2}).

The concept of 0 as a number comparable to the other basic digits is essential to this notation, as is the concept of 0's use as a placeholder, and as is the definition of multiplication and addition with 0. The use of 0 as a placeholder and, therefore, the use of a positional notation is first attested to in the Jain text from India entitled the *Lokavibhâga*, dated 458 AD and it was only in the early 13th century that these concepts, transmitted via the scholarship of the Arabic world, were introduced into Europe by Fibonacci using the Hindu–Arabic numeral system.

Algorism comprises all of the rules for performing arithmetic computations using this type of written numeral. For example, addition produces the sum of two arbitrary numbers. The result is calculated by the repeated addition of single digits from each number that occupies the same position, proceeding from right to left. An addition table with ten rows and ten columns displays all possible values for each sum. If an individual sum exceeds the value 9, the result is represented with two digits. The rightmost digit is the value for the current position, and the result for the subsequent addition of the digits to the left increases by the value of the second (leftmost) digit, which is always one. This adjustment is termed a *carry* of the value 1. The process for multiplying two arbitrary numbers is similar to the process for addition. A multiplication table with ten rows and ten columns lists the results for each pair of digits. If an individual product of a pair of digits exceeds 9, the *carry* adjustment increases the result of any subsequent multiplication from digits to the left by a value equal to the second (leftmost) digit, which is any value from 1 to 8 ($9 \times 9 = 81$). Additional steps define the final result.

Similar techniques exist for subtraction and division.

The creation of a correct process for multiplication relies on the relationship between values of adjacent digits. The value for any single digit in a numeral depends on its position. Also, each position to the left represents a value ten times larger than the position to the right. In mathematical terms, the exponent for the radix (base) of 10 increases by 1 (to the left) or decreases by 1 (to the right). Therefore, the value for any arbitrary digit is multiplied by a value of the form 10^n with integer n. The list of values corresponding to all possible positions for a single digit is written as $\{..., 10^2, 10, 1, 10^{-1}, 10^{-2}, ...\}$.

Repeated multiplication of any value in this list by 10 produces another value in the list. In mathematical terminology, this characteristic is defined as closure, and the previous list is described as **closed under multiplication**. It is the basis for correctly finding the results of multiplication using the previous technique. This outcome is one example of the uses of number theory.

Compound Unit Arithmetic

Compound unit arithmetic is the application of arithmetic operations to mixed radix quantities such as feet and inches, gallons and pints, pounds shillings and pence, and so on. Prior to the use of decimal-based systems of money and units of measure, the use of compound unit arithmetic formed a significant part of commerce and industry.

Basic Arithmetic Operations

The techniques used for compound unit arithmetic were developed over many centuries and are well-documented in many textbooks in many different languages. In addition to the basic arithmetic functions encountered in decimal arithmetic, compound unit arithmetic employs three more functions:

- **Reduction** where a compound quantity is reduced to a single quantity, for example conversion of a distance expressed in yards, feet and inches to one expressed in inches.

- **Expansion**, the inverse function to reduction, is the conversion of a quantity that is expressed as a single unit of measure to a compound unit, such as expanding 24 oz to 1 lb, 8 oz.
- **Normalization** is the conversion of a set of compound units to a standard form – for example rewriting "1 ft 13 in" as "2 ft 1 in".

Knowledge of the relationship between the various units of measure, their multiples and their submultiples forms an essential part of compound unit arithmetic.

Principles of Compound Unit Arithmetic

There are two basic approaches to compound unit arithmetic:

- **Reduction–expansion method** where all the compound unit variables are reduced to single unit variables, the calculation performed and the result expanded back to compound units. This approach is suited for automated calculations. A typical example is the handling of time by Microsoft Excel where all time intervals are processed internally as days and decimal fractions of a day.
- **On-going normalization method** in which each unit is treated separately and the problem is continuously normalized as the solution develops. This approach, which is widely described in classical texts, is best suited for manual calculations. An example of the ongoing normalization method as applied to addition is shown below.

The addition operation is carried out from right to left; in this case, pence are processed first, then shillings followed by pounds. The numbers below the "answer line" are intermediate results.

The total in the pence column is 25. Since there are 12 pennies in a shilling, 25 is divided by 12 to give 2 with a remainder of 1. The value "1" is then written to the answer row and the value "2" carried forward to the shillings column. This operation is repeated using the values in the shillings column, with the additional step of adding the value that was carried forward from the pennies column. The intermediate total is divided by 20 as there are 20 shillings in a pound. The pound column is then processed, but as pounds are the largest unit that is being considered, no values are carried forward from the pounds column.

It should be noted that for the sake of simplicity, the example chosen did not have farthings.

	£	s	d
	4	3	8
	5	17	7
	3	6	10
Answer	13	8	1
Column total	12	26	25
Carried forward	1	2	-
Unnormalised total	13	28	25
Normalisation	-	28÷20	25÷12

Operations in Practice

A scale calibrated in imperial units with an associated cost display.

During the 19th and 20th centuries various aids were developed to aid the manipulation of compound units, particularly in commercial applications. The most common aids were mechanical tills which were adapted in countries such as the United Kingdom to accommodate pounds, shillings, pennies and farthings and "Ready Reckoners" – books aimed at traders that catalogued the results of various routine calculations such as the percentages or multiples of various sums of money. One typical booklet that ran to 150 pages tabulated multiples "from one to ten thousand at the various prices from one farthing to one pound".

The cumbersome nature of compound unit arithmetic has been recognized for many years – in 1586, the Flemish mathematician Simon Stevin published a small pamphlet called *De Thiende* ("the tenth") in which he declared that the universal introduction of decimal coinage, measures, and weights to be merely a question of time while in the modern era, many conversion programs, such as that embedded in the calculator supplied as a standard part of the Microsoft Windows 7 operating system display compound units in a reduced decimal format rather than using an expanded format (i.e. "2.5 ft" is displayed rather than "2 ft 6 in").

Number Theory

Until the 19th century, *number theory* was a synonym of "arithmetic". The addressed problems were directly related to the basic operations and concerned primality, divisibility, and the solution of equations in integers, such as Fermat's last theorem. It appeared that most of these problems, although very elementary to state, are very difficult and may not be solved without very deep mathematics involving concepts and methods from many other branches of mathematics. This led to new branches of number theory such as analytic number theory, algebraic number theory, Diophantine geometry and arithmetic algebraic geometry. Wiles' proof of Fermat's Last Theorem is a typical example of the necessity of sophisticated methods, which go far beyond the classical methods of arithmetic, for solving problems that can be stated in elementary arithmetic.

Arithmetic in Education

Primary education in mathematics often places a strong focus on algorithms for the arithmetic of natural numbers, integers, fractions, and decimals (using the decimal place-value system). This study is sometimes known as algorism.

The difficulty and unmotivated appearance of these algorithms has long led educators to question this curriculum, advocating the early teaching of more central and intuitive mathematical ideas. One notable movement in this direction was the New Math of the 1960s and 1970s, which attempted to teach arithmetic in the spirit of axiomatic development from set theory, an echo of the prevailing trend in higher mathematics.

Also, arithmetic was used by Islamic Scholars in order to teach application of the rulings related to Zakat and Irth. This was done in a book entitled *The Best of Arithmetic* by Abd-al-Fattah-al-Dumyati.

The book begins with the foundations of mathematics and proceeds to its application in the later chapters.

REFERENCES

1. "Mathematics". Science Clarified. Retrieved 23 October 2012.
2. Davenport, Harold, The Higher Arithmetic: An Introduction to the Theory of Numbers (7th ed.), Cambridge University Press, Cambridge, UK, 1999, ISBN 0-521-63446-6.
3. Rudman, Peter Strom (2007). How Mathematics Happened: The First 50,000 Years. Prometheus Books. p. 64. ISBN 978-1-59102-477-4.
4. The Works of Archimedes, Chapter IV, Arithmetic in Archimedes, edited by T.L. Heath, Dover Publications Inc, New York, 2002.
5. Joseph Needham, Science and Civilization in China, Vol. 3, page 9, Cambridge University Press, 1959.
6. Reference: Revue de l'Orient Chretien by François Nau pp.327-338. (1929)
7. Reference: Sigler, L., "Fibonacci's Liber Abaci", Springer, 2003.
8. Tapson, Frank (1996). The Oxford Mathematics Study Dictionary. Oxford University Press. ISBN 0 19 914551 2.

9. Leonardo Pisano - page 3: "Contributions to number theory". Encyclopædia Britannica Online, 2006. Retrieved 18 September 2006.
10. Walkingame, Francis (1860). "The Tutor's Companion; or, Complete Practical Arithmetic" (PDF). Webb, Millington & Co. pp. 24–39.
11. Palaiseau, JFG (October 1816). Métrologie universelle, ancienne et moderne: ou rapport des poids et mesures des empires, royaumes, duchés et principautés des quatre parties du monde [Universal, ancient and modern metrology: or report of weights and measurements of empires, kingdoms, duchies and principalities of all parts of the world] (in French). Bordeaux. Retrieved October 30, 2011.
12. Jacob de Gelder (1824). Allereerste Gronden der Cijferkunst [Introduction to Numeracy] (in Dutch). 's-Gravenhage and Amsterdam: de Gebroeders van Cleef. pp. 163–176. Retrieved March 2, 2011.
13. Malaisé, Ferdinand (1842). Theoretisch-Praktischer Unterricht im Rechnen für die niederen Classen der Regimentsschulen der Königl. Bayer. Infantrie und Cavalerie[Theoretical and practical instruction in arithmetic for the lower classes of the Royal Bavarian Infantry and Cavalry School] (in German). Munich. Retrieved 20 March 2012.
14. Encyclopædia Britannica, Vol I, Edinburgh, 1772, Arithmetick
15. Walkingame, Francis (1860). "The Tutor's Companion; or, Complete Practical Arithmetic" (PDF). Webb, Millington & Co. pp. 43–50.
16. Thomson, J (1824). The Ready Reckoner in miniature containing accurate table from one to the thousand at the various prices from one farthing to one pound. Montreal. Retrieved 25 March 2012.
17. O'Connor, John J.; Robertson, Edmund F. (January 2004), "Arithmetic", MacTutor History of Mathematics archive, University of St Andrews.
18. Mathematically Correct: Glossary of Terms
19. al-Dumyati, Abd-al-Fattah Bin Abd-al-Rahman al-Banna (1887). "The Best of Arithmetic". World Digital Library (in Arabic). Retrieved 30 June 2013.
20. Blake, A. G. (1887). Arithmetic. Dublin, Ireland: Alexander Thom & Company.
21. Weisstein, Eric W. "Division". MathWorld.
22. Derbyshire, John (2004). Prime Obsession: Bernhard Riemann and the Greatest Unsolved Problem in Mathematics. New York City: Penguin Books. ISBN 978-0452285255.
23. Weisstein, Eric W. "Division by Zero". MathWorld.
24. Weisstein, Eric W. "Integer Division". MathWorld.
25. Earliest Uses of Symbols of Operation, Jeff MIller
26. Schmid, Hermann (1974). Decimal Computation (1 ed.). Binghamton, New York, USA: John Wiley & Sons. ISBN 0-471-76180-X.
27. Schmid, Hermann (1983) [1974]. Decimal Computation (1 (reprint) ed.). Malabar, Florida, USA: Robert E. Krieger Publishing Company. ISBN 0-89874-318-4.

28. "Subtraction". Oxford English Dictionary (3rd ed.). Oxford University Press. September 2005. (Subscription or UK public library membership required.)
29. "Subtrahend" is not a Latin word; in Latin it must be further conjugated, as in numerus subtrahendus "the number to be subtracted".
30. Paul E. Peterson, Michael Henderson, Martin R. West (2014) Teachers Versus the Public: What Americans Think about Schools and How to Fix Them Brookings Institution Press, p.163
31. Janet Kolodzy (2006) Convergence Journalism: Writing and Reporting across the News Media Rowman & Littlefield Publishers, p.180
32. David Gillborn (2008) Racism and Education: Coincidence Or Conspiracy? Routledge p.46
33. Paul Klapper (1916). The Teaching of Arithmetic: A Manual for Teachers. pp. 80–.
34. Susan Ross and Mary Pratt-Cotter. 2000. "Subtraction in the United States: An Historical Perspective," The Mathematics Educator 8(1):4-11. P. 8: "This new version of the decomposition algorithm [i.e., using Brownell's crutch] has so completely dominated the field that it is rare to see any other algorithm used to teach subtraction today [in America]."
35. Ross, Susan C.; Pratt-Cotter, Mary (1999). "Subtraction From a Historical Perspective". School Science and Mathematics. 99 (7): 389–393.
36. Klapper 1916, p. 177-.
37. David Eugene Smith (1913). The Teaching of Arithmetic. Ginn. pp. 77–.
38. The Many Ways of Arithmetic in UCSMP Everyday Mathematics Subtraction: Trade First
39. Partial-Differences Subtraction; The Many Ways of Arithmetic in UCSMP Everyday Mathematics Subtraction: Partial Differences
40. The Many Ways of Arithmetic in UCSMP Everyday Mathematics Subtraction: Counting Up
41. The Many Ways of Arithmetic in UCSMP Everyday Mathematics Subtraction: Left to Right Subtraction
42. The Many Ways of Arithmetic in UCSMP Everyday Mathematics Subtraction: Same Change Rule

CHAPTER 2

Fractions, Decimal and Percentage

Chapter Contents

Fractions give some people nightmares, but this doesn't have to be you. Keep watching this video lesson, and you will come out with a better understanding of fractions. And hopefully, you won't feel so afraid of them.

We begin with a definition of what fractions are. A fraction simply tells us how many parts of a whole we have. You can recognize a fraction by the slash that is written between the two numbers. We have a top number, the numerator, and a bottom number, the denominator. For example, 1/2 is a fraction. You can write it with a slanted slash like we have or you can write the 1 on top of the 2 with the slash between the two numbers. The 1 is the numerator, and the 2 is the denominator.

INTRODUCTION

A fraction (from Latin *fractus*, "broken") represents a part of a whole or, more generally, any number of equal parts. When spoken in everyday English, a fraction describes how many parts of a certain size there are, for example, one-half, eight-fifths, three-quarters. A *common*, *vulgar*, or *simple* fraction (examples: $\frac{1}{2}$ and 17/3) consists of an integer numerator displayed above a line (or before a slash), and a non-zero integer denominator, displayed below (or after) that line. Numerators and denominators are also used in fractions that are not *common*, including compound fractions, complex fractions, and mixed numerals.

The numerator represents a number of equal parts, and the denominator, which cannot be zero, indicates how many of those parts make up a unit or a whole. For example, in the fraction 3/4, the numerator, 3, tells us that the fraction represents 3 equal parts, and the denominator, 4, tells us that 4 parts make up a whole. The picture to the right illustrates $\frac{3}{4}$ or ¾ of a cake.

Fractional numbers can also be written without using explicit numerators or denominators, by using decimals, percent signs, or negative exponents (as in 0.01, 1%, and 10^{-2} respectively, all of which are equivalent to 1/100). An integer such as the number 7 can be thought of as having an implicit denominator of one: 7 equals 7/1.

Other uses for fractions are to represent ratios and to represent division. Thus the fraction ¾ is also used to represent the ratio 3:4 (the ratio of the part to the whole) and the division 3 ÷ 4 (three divided by four).

In mathematics the set of all numbers that can be expressed in the form a/b, where a and b are integers and b is not zero, is called the set of rational numbers and is represented by the symbol **Q**, which stands for quotient. The test for a number being a rational number is that it can be written in that form (i.e., as a common fraction). However, the word *fraction* is also used to describe mathematical expressions that are not rational numbers, for example algebraic fractions (quotients of algebraic expressions), and expressions that contain irrational numbers, such as $\sqrt{2}/2$ (see square root of 2) and $\pi/4$ (see proof that π is irrational).

A cake with one quarter (one fourth) removed. The remaining three fourths are shown. Dotted lines indicate where the cake may be cut in order to divide it into equal parts. Each fourth of the cake is denoted by the fraction ¼.

Vocabulary

In a fraction, the number of equal parts being described is the numerator (equivalent to the dividend in division) and the number of equal parts that make up a whole is the denominator (equivalent to the divisor). Informally, they may be distinguished by placement alone but in formal contexts they are always separated by a fraction bar. The fraction bar may be horizontal (as in −1/3), oblique (as in 1/5 or −1⁄7), or diagonal (as in ¹⁄₉). These marks are respectively known as the horizontal bar, the slash (US) or stroke (UK), the division slash, and the fraction slash.[n 1] In typography, horizontal fractions are also known as "en" or "nut fractions" and diagonal fractions as "em fractions", based on the width of a line they take up.

The denominators of English fractions are generally expressed as ordinal numbers, in the plural if the numerator is not one. (For example, ²⁄₅ and ³⁄₅ are both read as a number of "fifths".) Exceptions include the denominator 2, which is always read "half" or "halves", the denominator 4, which may be alternatively expressed as "quarter"/"quarters" or as "fourth"/"fourths", and the denominator 100, which may be alternatively expressed as "hundredth"/"hundredths" or "percent". When the denominator is 1, it may be expressed in terms of "wholes" but is more commonly ignored, with the numerator read out as a whole number. (For example, 3/1 may be described as "three wholes" or as simply "three".) When the numerator is one, it may be omitted. (For example, "a tenth" or "each quarter".) A fraction may be expressed as a single composition, in which case it is hyphenated, or as a number of fractions with a numerator of one, in which case they are not. (For example, "two-fifths" is the fraction 2/5 and "two fifths" is the same fraction understood as 2 instances of ¹⁄₅.) Fractions should always be hyphenated when used as adjectives.

Alternatively, a fraction may be described by reading it out as the numerator "over" the denominator, with the denominator expressed as a cardinal number. (For example, 3/1 may also be expressed as "three over one".) The term "over" is used even in the case of solidus fractions, where the numbers are placed left and right of a slash mark. (For example, ½ may be read "one-half", "one half", or "one over two".) Fractions with large denominators that are *not* powers of ten are often rendered in this fashion (e.g., 1/117 as "one over a hundred and seventeen") while those with denominators divisible by ten are typically read in the normal ordinal fashion (e.g., 6/1000000 as "six-millionths", "six millionths", or "six one-millionths").

Forms of Fractions

Simple, Common, or Vulgar Fractions

A simple fraction (also known as a common fraction or vulgar fraction) is a rational number written as *a*/*b* or $\frac{a}{b}$ where *a* and *b* are both integers. As with other fractions, the denominator (*b*) cannot be zero. Examples include $\frac{1}{2}$, $-\frac{8}{5}$, $\frac{-8}{5}$, $\frac{8}{-5}$, and 3/17. *Simple fractions* can be positive or negative, proper, or improper (see below). Compound fractions, complex fractions, mixed numerals, and decimals (see below) are not *simple fractions*, though, unless irrational, they can be evaluated to a simple fraction.

Proper and Improper Fractions

Common fractions can be classified as either proper or improper. When the numerator and the denominator are both positive, the fraction is called proper if the numerator is less than the denominator, and improper otherwise. In general, a common fraction is said to be a proper fraction if the absolute value of the fraction is strictly less than one—that is, if the fraction is greater than −1 and less than 1. It is said to be an improper fraction, or sometimes top-heavy fraction, if the absolute value of the fraction is greater than or equal to 1. Examples of proper fractions are 2/3, –3/4, and 4/9; examples of improper fractions are 9/4, –4/3, and 3/3.

Mixed Numbers

A **mixed numeral** (also called a *mixed number* or *mixed fraction*) is the sum of a non-zero integer and a proper fraction. This sum is implied without the use of any visible operator such as "+". For example, in referring to two entire cakes and three quarters of another cake, the whole and fractional parts of the number are written next to each other $2+\frac{3}{4}=2\frac{3}{4}$.

This is not to be confused with the algebra rule of implied multiplication. When two algebraic expressions are written next to each other, the operation of multiplication is said to be "understood". In algebra, $a\frac{b}{c}$ for example is not a mixed number. Instead, multiplication is understood where $a\frac{b}{c} = a \times \frac{b}{c}$.

To avoid confusion, the multiplication is often explicitly expressed. So $a\frac{b}{c}$ may be written as

$$a \times \frac{b}{c},$$

$$a \cdot \frac{b}{c},$$

or

$$a\left(\frac{b}{c}\right).$$

An improper fraction is another way to write a whole plus a part. A mixed number can be converted to an improper fraction as follows:

1. Write the mixed number $2\frac{3}{4}$ as a sum $2+\frac{3}{4}$.
2. Convert the whole number to an improper fraction with the same denominator as the fractional part, $2=\frac{8}{4}$.
3. Add the fractions. The resulting sum is the improper fraction. In the example, $2\frac{3}{4}=\frac{8}{4}+\frac{3}{4}=\frac{11}{4}$.

Similarly, an improper fraction can be converted to a mixed number as follows:

- Divide the numerator by the denominator. In the example, 11/4 divide 11 by 4. 11 ÷ 4 = 2 with remainder 3
- The quotient (without the remainder) becomes the whole number part of the mixed number. The remainder becomes the numerator of the fractional part. In the example, 2 is the whole number part and 3 is the numerator of the fractional part.
- The new denominator is the same as the denominator of the improper fraction. In the example, they are both 4. Thus $\frac{11}{4} = 2\frac{3}{4}$.

Mixed numbers can also be negative, as in $-2\frac{3}{4}$, which equals $-(2 + \frac{3}{4}) = -2 - \frac{3}{4}$.

Ratios

A ratio is a relationship between two or more numbers that can be sometimes expressed as a fraction. Typically, a number of items are grouped and compared in a ratio, specifying numerically the relationship between each group. Ratios are expressed as "group 1 to group 2 ... to group *n*". For example, if a car lot had 12 vehicles, of which

- 2 are white,
- 6 are red, and
- 4 are yellow,

then the ratio of red to white to yellow cars is 6 to 2 to 4. The ratio of yellow cars to white cars is 4 to 2 and may be expressed as 4:2 or 2:1.

A ratio is often converted to a fraction when it is expressed as a ratio to the whole. In the above example, the ratio of yellow cars to all the cars on the lot is 4:12 or 1:3. We can convert these ratios to a fraction and say that 4/12 of the cars or ⅓ of the cars in the lot are yellow. Therefore, if a person randomly chose one car on the lot, then there is a one in three chance or probability that it would be yellow.

Reciprocals and the "Invisible Denominator"

The **reciprocal** of a fraction is another fraction with the numerator and denominator exchanged. The reciprocal of $\frac{3}{7}$, for instance, is $\frac{7}{3}$. The product of a fraction and its reciprocal is 1, hence the reciprocal is the multiplicative inverse of a fraction.

Any integer can be written as a fraction with the number one as denominator. For example, 17 can be

written as $\frac{17}{1}$, where 1 is sometimes referred to as the *invisible denominator*. Therefore, every fraction or integer except for zero has a reciprocal. The reciprocal of 17 is $\frac{1}{17}$.

Complex Fractions

In a **complex fraction**, either the numerator, or the denominator, or both, is a fraction or a mixed number, corresponding to division of fractions. For example, $\frac{\frac{1}{2}}{\frac{1}{3}}$ and $\frac{12\frac{3}{4}}{26}$ are complex fractions. To reduce a complex fraction to a simple fraction, treat the longest fraction line as representing division. For example:

$$\frac{\frac{1}{2}}{\frac{1}{3}} = \frac{1}{2} \times \frac{3}{1} = \frac{3}{2} = 1\frac{1}{2}$$

$$\frac{12\frac{3}{4}}{26} = 12\frac{3}{4} \cdot \frac{1}{26} = \frac{12 \cdot 4+3}{4} \cdot \frac{1}{26} = \frac{51}{4} \cdot \frac{1}{26} = \frac{51}{104}$$

$$\frac{\frac{3}{2}}{5} = \frac{3}{2} \times \frac{1}{5} = \frac{3}{10}$$

$$\frac{8}{\frac{1}{3}} = 8 \times \frac{3}{1} = 24.$$

If, in a complex fraction, there is no clear way to tell which fraction lines takes precedence, then the expression is obviously improperly formed, because of ambiguity. So 5/10/20/40 is a poorly constructed mathematical expression, with multiple possible meanings, e.g.:

$$\frac{5}{10/\frac{20}{40}} = \frac{1}{4} \quad \text{or} \quad \frac{\frac{5}{10}}{\frac{20}{40}} = 1$$

Compound Fractions

A compound fraction is a fraction of a fraction, or any number of fractions connected with the word *of*, corresponding to multiplication of fractions. To reduce a compound fraction to a simple fraction, just carry out the multiplication (see the section on multiplication). For example, $\frac{3}{4}$ of $\frac{5}{7}$ is a compound fraction, corresponding to $\frac{3}{4} \times \frac{5}{7} = \frac{15}{28}$. The terms compound fraction and complex fraction are closely related and sometimes one is used as a synonym for the other.

Decimal Fractions and Percentages

A decimal fraction is a fraction whose denominator is not given explicitly, but is understood to be an integer power of ten. Decimal fractions are commonly expressed using decimal notation in which the implied denominator is determined by the number of digits to the right of a decimal separator, the appearance of which (e.g., a period, a raised period (•), a comma) depends on the locale (for examples, see decimal separator). Thus for 0.75 the numerator is 75 and the implied denominator is 10 to the second power, *viz.* 100, because there are two digits to the right of the decimal separator. In decimal numbers greater than 1 (such as 3.75), the fractional part of the number is expressed by the digits to the right of the decimal (with a value of 0.75 in this case). 3.75 can be written either as an improper fraction, 375/100, or as a mixed number, $3\frac{75}{100}$.

Decimal fractions can also be expressed using scientific notation with negative exponents, such as 6.023×10^{-7}, which represents 0.0000006023. The 10^{-7} represents a denominator of 10^7. Dividing by 10^7 moves the decimal point 7 places to the left.

Decimal fractions with infinitely many digits to the right of the decimal separator represent an infinite series. For example, ⅓ = 0.333... represents the infinite series 3/10 + 3/100 + 3/1000 + ...

Another kind of fraction is the percentage (Latin *per centum* meaning "per hundred", represented by the symbol %), in which the implied denominator is always 100. Thus, 51% means 51/100. Percentages greater than 100 or less than zero are treated in the same way, e.g. 311% equals 311/100, and −27% equals −27/100.

The related concept of *permille* or *parts per thousand* has an implied denominator of 1000, while the more general parts-per notation, as in 75 parts per million, means that the proportion is 75/1,000,000.

Whether common fractions or decimal fractions are used is often a matter of taste and context. Common fractions are used most often when the denominator is relatively small. By mental calculation, it is easier to multiply 16 by 3/16 than to do the same calculation using the fraction's decimal equivalent (0.1875). And it is more accurate to multiply 15 by 1/3, for example, than it is to multiply 15 by any decimal approximation of one third. Monetary values are commonly expressed as decimal fractions, for example $3.75. However, as noted above, in pre-decimal British currency, shillings and pence were often given the form (but not the meaning) of a fraction, as, for example 3/6 (read "three and six") meaning 3 shillings and 6 pence, and having no relationship to the fraction 3/6.

Special Cases

- A unit fraction is a vulgar fraction with a numerator of 1, e.g. 1/7. Unit fractions can also be expressed using negative exponents, as in 2^{-1}, which represents 1/2, and 2^{-2}, which represents $1/(2^2)$ or 1/4.
- An Egyptian fraction is the sum of distinct positive unit fractions, for example $\frac{1}{2}+\frac{1}{3}$ This definition derives from the fact that the ancient Egyptians expressed all fractions except $\frac{1}{2}$, $\frac{2}{3}$ and $\frac{3}{4}$ in this manner. Every positive rational number can be expanded as an Egyptian fraction. For example, $\frac{5}{7}$ can be written as

$\frac{1}{2}+\frac{1}{6}+\frac{1}{21}$. Any positive rational number can be written as a sum of unit fractions in infinitely many ways. Two ways to write $\frac{13}{17}$ are $\frac{1}{2}+\frac{1}{4}+\frac{1}{68}$ and $\frac{1}{3}+\frac{1}{4}+\frac{1}{6}+\frac{1}{68}$.

- A dyadic fraction is a vulgar fraction in which the denominator is a power of two, e.g. $\frac{1}{8}$

Arithmetic with Fractions

Like whole numbers, fractions obey the commutative, associative, and distributive laws, and the rule against division by zero.

Equivalent Fractions

Multiplying the numerator and denominator of a fraction by the same (non-zero) number results in a fraction that is equivalent to the original fraction. This is true because for any non-zero number n, the fraction $\frac{n}{n}=1$. Therefore, multiplying by $\frac{n}{n}$ is equivalent to multiplying by one, and any number multiplied by one has the same value as the original number. By way of an example, start with the fraction $\frac{1}{2}$. When the numerator and denominator are both multiplied by 2, the result is $\frac{2}{4}$, which has the same value (0.5) as $\frac{1}{2}$. To picture this visually, imagine cutting a cake into four pieces; two of the pieces together ($\frac{2}{4}$) make up half the cake ($\frac{1}{2}$).

Dividing the numerator and denominator of a fraction by the same non-zero number will also yield an equivalent fraction. This is called reducing or simplifying the fraction. A simple fraction in which the numerator and denominator are coprime (that is, the only positive integer that goes into both the numerator and denominator evenly is 1) is said to be irreducible, in lowest terms, or in simplest terms.

For example, $\frac{3}{9}$ is not in lowest terms because both 3 and 9 can be exactly divided by 3. In contrast, $\frac{3}{8}$ *is* in lowest terms—the only positive integer that goes into both 3 and 8 evenly is 1.

Using these rules, we can show that $\frac{5}{10}=\frac{1}{2}=\frac{10}{20}=\frac{50}{100}$.

A common fraction can be reduced to lowest terms by dividing both the numerator and denominator by their greatest common divisor. For example, as the greatest common divisor of 63 and 462 is 21, the fraction $\frac{63}{462}$ can be reduced to lowest terms by dividing the numerator and denominator by 21:

$$\frac{63}{462}=\frac{63\div 21}{462\div 21}=\frac{3}{22}$$

The Euclidean algorithm gives a method for finding the greatest common divisor of any two positive integers.

Comparing Fractions

Comparing fractions with the same denominator only requires comparing the numerators.

$\frac{3}{4} > \frac{2}{4}$ because 3>2.

If two positive fractions have the same numerator, then the fraction with the smaller denominator is the larger number. When a whole is divided into equal pieces, if fewer equal pieces are needed to make up the whole, then each piece must be larger. When two positive fractions have the same numerator, they represent the same number of parts, but in the fraction with the smaller denominator, the parts are larger.

One way to compare fractions with different numerators and denominators is to find a common denominator. To compare $\frac{a}{b}$ and $\frac{c}{d}$, these are converted to $\frac{ad}{bd}$ and $\frac{bc}{bd}$. Then *bd* is a common denominator and the numerators *ad* and *bc* can be compared.

$\frac{2}{3} ? \frac{1}{2}$ gives $\frac{4}{6} > \frac{3}{6}$

It is not necessary to determine the value of the common denominator to compare fractions. This short cut is known as "cross multiplying" – you can just compare *ad* and *bc*, without computing the denominator.

$\frac{5}{18} ? \frac{4}{17}$

Multiply top and bottom of each fraction by the denominator of the other fraction, to get a common denominator:

$\frac{5\times17}{18\times17} ? \frac{4\times18}{17\times18}$

The denominators are now the same, but it is not necessary to calculate their value – only the numerators need to be compared. Since 5×17 (= 85) is greater than 4×18 (= 72), $\frac{5}{18} > \frac{4}{17}$.

Also note that every negative number, including negative fractions, is less than zero, and every positive number, including positive fractions, is greater than zero, so every negative fraction is less than any positive fraction.

Addition

The first rule of addition is that only like quantities can be added; for example, various quantities of quarters. Unlike quantities, such as adding thirds to quarters, must first be converted to like quantities as described below: Imagine a pocket containing two quarters, and another pocket containing three quarters; in total, there are five quarters. Since four quarters is equivalent to one (dollar), this can be represented as follows:

$$\frac{2}{4} + \frac{3}{4} = \frac{5}{4} = 1\frac{1}{4}.$$

Adding Unlike Quantities

To add fractions containing unlike quantities (e.g. quarters and thirds), it is necessary to convert all amounts to like quantities. It is easy to work out the chosen type of fraction to convert to; simply multiply together the two denominators (bottom number) of each fraction.

For adding quarters to thirds, both types of fraction are converted to twelfths, thus:

If $\frac{1}{2}$ of a cake is to be added to $\frac{1}{4}$ of a cake, the pieces need to be converted into comparable quantities, such as cake-eighths or cake-quarters.

$$\frac{1}{4}+\frac{1}{3}=\frac{1\times 3}{4\times 3}+\frac{1\times 4}{3\times 4}=\frac{3}{12}+\frac{4}{12}=\frac{7}{12}.$$

Consider adding the following two quantities:

$$\frac{3}{5}+\frac{2}{3}$$

First, convert $\frac{3}{5}$ into fifteenths by multiplying both the numerator and denominator by three:

$\frac{3}{5}\times\frac{3}{3}=\frac{9}{15}$. Since $\frac{3}{3}$ equals 1, multiplication by $\frac{3}{3}$ does not change the value of the fraction.

Second, convert $\frac{2}{3}$ into fifteenths by multiplying both the numerator and denominator by five:

$\frac{2}{3}\times\frac{5}{5}=\frac{10}{15}$.

Now it can be seen that:

$$\frac{3}{5}+\frac{2}{3}$$

is equivalent to:

$$\frac{9}{15}+\frac{10}{15}=\frac{19}{15}=1\frac{4}{15}$$

This method can be expressed algebraically:

$$\frac{a}{b}+\frac{c}{d}=\frac{ad+cb}{bd}$$

And for expressions consisting of the addition of three fractions:

$$\frac{a}{b}+\frac{c}{d}+\frac{e}{f}=\frac{a(df)+c(bf)+e(bd)}{bdf}$$

This method always works, but sometimes there is a smaller denominator that can be used (a least common denominator). For example, to add $\frac{3}{4}$ and $\frac{5}{12}$ the denominator 48 can be used (the product of 4 and 12), but the smaller denominator 12 may also be used, being the least common multiple of 4 and 12.

$$\frac{3}{4}+\frac{5}{12}=\frac{9}{12}+\frac{5}{12}=\frac{14}{12}=\frac{7}{6}=1\frac{1}{6}$$

Subtraction

The process for subtracting fractions is, in essence, the same as that of adding them: find a common denominator, and change each fraction to an equivalent fraction with the chosen common denominator. The resulting fraction will have that denominator, and its numerator will be the result of subtracting the numerators of the original fractions. For instance,

$$\frac{2}{3} - \frac{1}{2} = \frac{4}{6} - \frac{3}{6} = \frac{1}{6}$$

Multiplication

Multiplying a Fraction by Another Fraction

To multiply fractions, multiply the numerators and multiply the denominators. Thus:

$$\frac{2}{3} \times \frac{3}{4} = \frac{6}{12}$$

To explain the process, consider one third of one quarter. Using the example of a cake, if three small slices of equal size make up a quarter, and four quarters make up a whole, twelve of these small, equal slices make up a whole. Therefore, a third of a quarter is a twelfth. Now consider the numerators. The first fraction, two thirds, is twice as large as one third. Since one third of a quarter is one twelfth, two thirds of a quarter is two twelfth. The second fraction, three quarters, is three times as large as one quarter, so two thirds of three quarters is three times as large as two thirds of one quarter. Thus two thirds times three quarters is six twelfths.

A short cut for multiplying fractions is called "cancellation". Effectively the answer is reduced to lowest terms during multiplication. For example:

$$\frac{2}{3} \times \frac{3}{4} = \frac{\cancel{2}^{\,1}}{\cancel{3}_{\,1}} \times \frac{\cancel{3}^{\,1}}{\cancel{4}_{\,2}} = \frac{1}{1} \times \frac{1}{2} = \frac{1}{2}$$

A two is a common factor in both the numerator of the left fraction and the denominator of the right and is divided out of both. Three is a common factor of the left denominator and right numerator and is divided out of both.

Multiplying a Fraction by a Whole Number

Since a whole number can be rewritten as itself divided by 1, normal fraction multiplication rules can still apply.

$6 \times \frac{3}{4} = \frac{6}{1} \times \frac{3}{4} = \frac{18}{4}$ This method works because the fraction 6/1 means six equal parts, each one of which is a whole.

Multiplying Mixed Numbers

When multiplying mixed numbers, it is considered preferable to convert the mixed number into an improper fraction. For example:

$$3 \times 2\tfrac{3}{4} = 3 \times \left(\tfrac{8}{4} + \tfrac{3}{4}\right) = 3 \times \tfrac{11}{4} = \tfrac{33}{4} = 8\tfrac{1}{4}$$

In other words, $2\frac{3}{4}$ is the same as $\frac{8}{4} + \frac{3}{4}$, making 11 quarters in total (because 2 cakes, each split into quarters makes 8 quarters total) and 33 quarters is $8\frac{1}{4}$, since 8 cakes, each made of quarters, is 32 quarters in total.

Division

To divide a fraction by a whole number, you may either divide the numerator by the number, if it goes evenly into the numerator, or multiply the denominator by the number. For example, $\frac{10}{3} \div 5$ equals $\frac{2}{3}$ and also equals $\frac{10}{3 \cdot 5} = \frac{10}{15}$, which reduces to $\frac{2}{3}$. To divide a number by a fraction, multiply that number by the reciprocal of that fraction. Thus, $\frac{1}{2} \div \frac{3}{4} = \frac{1}{2} \times \frac{4}{3} = \frac{1 \cdot 4}{2 \cdot 3} = \frac{2}{3}$.

Converting Between Decimals and Fractions

To change a common fraction to a decimal, divide the denominator into the numerator. Round the answer to the desired accuracy. For example, to change ¼ to a decimal, divide 4 into 1.00, to obtain 0.25. To change ⅓ to a decimal, divide 3 into 1.0000..., and stop when the desired accuracy is obtained. Note that ¼ can be written exactly with two decimal digits, while ⅓ cannot be written exactly with any finite number of decimal digits. To change a decimal to a fraction, write in the denominator a 1 followed by as many zeroes as there are digits to the right of the decimal point, and write in the numerator all the digits in the original decimal, omitting the decimal point. Thus 12.3456 = 123456/10000.

Converting Repeating Decimals to Fractions

Decimal numbers, while arguably more useful to work with when performing calculations, sometimes lack the precision that common fractions have. Sometimes an infinite repeating decimal is required to reach the same precision. Thus, it is often useful to convert repeating decimals into fractions.

The preferred way to indicate a repeating decimal is to place a bar over the digits that repeat, for example 0.789 = 0.789789789… For repeating patterns where the repeating pattern begins immediately after the decimal point, a simple division of the pattern by the same number of nines as numbers it has will suffice. For example:

$0.\overline{5} = 5/9$

$0.\overline{62} = 62/99$

$0.\overline{264} = 264/999$

$0.\overline{6291} = 6291/9999$

In case leading zeros precede the pattern, the nines are suffixed by the same number of trailing zeros:

$0.0\overline{5} = 5/90$

$0.000\overline{392} = 392/999000$

$0.00\overline{12} = 12/9900$

In case a non-repeating set of decimals precede the pattern (such as $0.1523\overline{987}$), we can write it as the sum of the non-repeating and repeating parts, respectively:

$0.1523 + 0.0000\overline{987}$

Then, convert both parts to fractions, and add them using the methods described above:

1523/10000 + 987/9990000 = 1522464/9990000

Alternatively, algebra can be used, such as below:

1. Let x = the repeating decimal:

   $x = 0.1523\overline{987}$

2. Multiply both sides by the power of 10 just great enough (in this case 10^4) to move the decimal point just before the repeating part of the decimal number:

   $10{,}000x = 1{,}523.\overline{987}$

3. Multiply both sides by the power of 10 (in this case 10^3) that is the same as the number of places that repeat:

   $10{,}000{,}000x = 1{,}523{,}987.\overline{987}$

4. Subtract the two equations from each other (if $a = b$ and $c = d$, then $a - c = b - d$):

   $10{,}000{,}000x - 10{,}000x = 1{,}523{,}987.\overline{987} - 1{,}523.\overline{987}$

5. Continue the subtraction operation to clear the repeating decimal:

   9,990,000x = 1,523,987 − 1,523

   9,990,000x = 1,522,464

6. Divide both sides to represent x as a fraction

   x = 1522464/9990000

Fractions in Abstract Mathematics

In addition to being of great practical importance, fractions are also studied by mathematicians, who check that the rules for fractions given above are consistent and reliable. Mathematicians define a fraction as an ordered pair (a, b) of integers a and $b \neq 0$, for which the operations addition, subtraction, multiplication, and division are defined as follows:

$$(a, b) + (c, d) = (ad + bc, bd)$$
$$(a, b) - (c, d) = (ad - bc, bd)$$
$$(a, b) \cdot (c, d) = (ac, bd)$$
$$(a, b) \div (c, d) = (ad, bc) \text{ (when c} \neq 0)$$

In addition, an equivalence relation is specified as follows: $(a, b) \sim (c, d)$ if and only if $ad = bc$.

These definitions agree in every case with the definitions given above; only the notation is different.

More generally, *a* and *b* may be elements of any integral domain *R*, in which case a fraction is an element of the field of fractions of *R*. For example, when *a* and *b* are polynomials in one indeterminate, the field of fractions is the field of rational fractions (also known as the field of rational functions). When *a* and *b* are integers, the field of fractions is the field of rational numbers.

Algebraic Fractions

An algebraic fraction is the indicated quotient of two algebraic expressions. Two examples of algebraic fractions are $\frac{3x}{x^2 + 2x - 3}$ and $\frac{\sqrt{x+2}}{x^2 - 3}$. Algebraic fractions are subject to the same laws as arithmetic fractions.

If the numerator and the denominator are polynomials, as in $\frac{3x}{x^2 + 2x - 3}$, the algebraic fraction is called a **rational fraction** (or **rational expression**). An irrational fraction is one that contains the variable under a fractional exponent or root, as in $\frac{\sqrt{x+2}}{x^2 - 3}$.

The terminology used to describe algebraic fractions is similar to that used for ordinary fractions. For example, an algebraic fraction is in lowest terms if the only factors common to the numerator and the denominator are 1 and −1. An algebraic fraction whose numerator or denominator, or both, contain a fraction, such as $\frac{1 + \frac{1}{x}}{1 - \frac{1}{x}}$ is called a **complex fraction**.

Rational numbers are the quotient field of integers. Rational expressions are the quotient field of the polynomials (over some integral domain). Since a coefficient is a polynomial of degree zero, a radical expression such as $\sqrt{2}/2$ is a rational fraction. Another example (over the reals) is $\frac{\pi}{2}$, the radian measure of a right angle.

The term partial fraction is used when decomposing rational expressions into sums. The goal is to write the rational expression as the sum of other rational expressions with denominators of lesser degree. For example, the rational expression $\frac{2x}{x^2-1}$ can be rewritten as the sum of two fractions: $\frac{1}{x+1} + \frac{1}{x-1}$ This is useful in many areas such as integral calculus and differential equations.

Radical Expressions

A fraction may also contain radicals in the numerator and/or the denominator. If the denominator contains radicals, it can be helpful to rationalize it (compare Simplified form of a radical expression), especially if further operations, such as adding or comparing that fraction to another, are to be carried out. It is also more convenient if division is to be done manually. When the denominator is a monomial square root, it can be rationalized by multiplying both the top and the bottom of the fraction by the denominator:

$$\frac{3}{\sqrt{7}} = \frac{3}{\sqrt{7}} \cdot \frac{\sqrt{7}}{\sqrt{7}} = \frac{3\sqrt{7}}{7}$$

The process of rationalization of binomial denominators involves multiplying the top and the bottom of a fraction by the conjugate of the denominator so that the denominator becomes a rational number. For example:

$$\frac{3}{3-2\sqrt{5}} = \frac{3}{3-2\sqrt{5}} \cdot \frac{3+2\sqrt{5}}{3+2\sqrt{5}} = \frac{3(3+2\sqrt{5})}{3^2-(2\sqrt{5})^2} = \frac{3(3+2\sqrt{5})}{9-20} = -\frac{9+6\sqrt{5}}{11}$$

$$\frac{3}{3+2\sqrt{5}} = \frac{3}{3+2\sqrt{5}} \cdot \frac{3-2\sqrt{5}}{3-2\sqrt{5}} = \frac{3(3-2\sqrt{5})}{3^2-(2\sqrt{5})^2} = \frac{3(3-2\sqrt{5})}{9-20} = -\frac{9-6\sqrt{5}}{11}$$

Even if this process results in the numerator being irrational, like in the examples above, the process may still facilitate subsequent manipulations by reducing the number of irrationals one has to work with in the denominator.

Typographical Variations

In computer displays and typography, simple fractions are sometimes printed as a single character, e.g. ½ (one half). See the article on Number Forms for information on doing this in Unicode.

Scientific publishing distinguishes four ways to set fractions, together with guidelines on use:

- **special fractions:** fractions that are presented as a single character with a slanted bar, with roughly the same height and width as other characters in the text. Generally used for simple fractions, such as: ½, ⅓, ⅔, ¼, and ¾. Since the numerals are smaller, legibility can be an issue, especially for small-sized fonts. These are not used in modern mathematical notation, but in other contexts.
- **case fractions:** similar to special fractions, these are rendered as a single typographical character, but with a horizontal bar, thus making them *upright*. An example would be $\frac{1}{2}$ but rendered with the same height as other characters. Some sources include all rendering of fractions as *case fractions* if they take only one typographical space, regardless of the direction of the bar.
- **shilling** or **solidus fractions:** 1/2, so called because this notation was used for pre-decimal British currency (£sd), as in 2/6 for a half crown, meaning two

shillings and six pence. While the notation "two shillings and six pence" did not represent a fraction, the forward slash is now used in fractions, especially for fractions inline with prose (rather than displayed), to avoid uneven lines. It is also used for fractions within fractions (complex fractions) or within exponents to increase legibility. Fractions written this way, also known as *piece fractions*, are written all on one typographical line, but take 3 or more typographical spaces.

built-up fractions: $\frac{1}{2}$ This notation uses two or more lines of ordinary text, and results in a variation in spacing between lines when included within other text. While large and legible, these can be disruptive, particularly for simple fractions or within complex fractions.

The earliest fractions were reciprocals of integers: ancient symbols representing one part of two, one part of three, one part of four, and so on. The Egyptians used Egyptian fractions c. 1000 bc. About 4000 years ago, Egyptians divided with fractions using slightly different methods. They used least common multiples with unit fractions. Their methods gave the same answer as modern methods. The Egyptians also had a different notation for dyadic fractions in the Akhmim Wooden Tablet and several Rhind Mathematical Papyrus problems.

The Greeks used unit fractions and (later) continued fractions. Followers of the Greek philosopherPythagoras (c. 530 bc) discovered that the square root of two cannot be expressed as a fraction of integers. (This is commonly though probably erroneously ascribed to Hippasus of Metapontum, who is said to have been executed for revealing this fact.) In 150 bc Jain mathematicians in India wrote the "Sthananga Sutra", which contains work on the theory of numbers, arithmetical operations, and operations with fractions.

The modern expression of fractions seems to have originated in India in the work of Aryabhatta (c. ad 500), Brahmagupta (c. 628), and Bhaskara (c. 1150). Their works form fractions by placing the numerators (Sanskrit: *amsa*) over the denominators (*cheda*), but without a bar between them. In Sanskrit literature, fractions were always expressed as an addition to or subtraction from an integer. The integer was written on one line and the fraction in its two parts on the next line. If the fraction was marked by a small circle ⟨∘⟩ or cross ⟨+⟩, it is subtracted from the integer; if no such sign appears, it is understood to be added. For example, Bhaskara I writes

६	१	२
१	१	१∘
४	५	९

which is the equivalent of

6	1	2
1	1	−1
4	5	9

and would be written in modern notation as $6\frac{1}{4}$, $1\frac{1}{5}$, and $2-\frac{1}{9}$ (i.e., $1\frac{8}{9}$).

The horizontal fraction bar is first attested in the work of Al-Hassār (fl. 1200), a Muslim mathematician from Fez, Morocco, who specialized in Islamic inheritance jurisprudence. In his discussion he writes, "... for example, if you are told to write three-fifths and a third of a fifth, write thus $\frac{3 \quad 1}{5 \quad 3}$ The same fractional notation—with the fraction given before the integer—appears soon after in the work of Leonardo Fibonacci in the 13th century.

In discussing the origins of decimal fractions, Dirk Jan Struik states:

The introduction of decimal fractions as a common computational practice can be dated back to the Flemish pamphlet *De Thiende*, published at Leyden in 1585, together with a French translation, *La Disme*, by the Flemish mathematician Simon Stevin (1548–1620), then settled in the Northern Netherlands. It is true that decimal fractions were used by the Chinese many centuries before Stevin and that the Persian astronomer Al-Kāshī used both decimal and sexagesimal fractions with great ease in his *Key to arithmetic* (Samarkand, early fifteenth century)."

While the Persian mathematician Jamshīd al-Kāshī claimed to have discovered decimal fractions himself in the 15th century, J. Lennart Berggren notes that he was mistaken, as decimal fractions were first used five centuries before him by the Baghdadi mathematician Abu'l-Hasan al-Uqlidisi as early as the 10th century.

In Formal Education

Pedagogical Tools

In primary schools, fractions have been demonstrated through Cuisenaire rods, Fraction Bars, fraction strips, fraction circles, paper (for folding or cutting), pattern blocks, pie-shaped pieces, plastic rectangles, grid paper, dot paper, geoboards, counters and computer software.

Documents for Teachers

Several states in the United States have adopted learning trajectories from the Common Core State Standards Initiative's guidelines for mathematics education. Aside from sequencing the learning of fractions and operations with fractions, the document provides the following definition of a fraction: "A number expressible in the form $\frac{a}{b}$ where a is a whole number and b is a positive whole number. (The word *fraction* in the standards always refers to a non-negative number.)" The document itself also refers to negative fractions.

DECIMAL

The **decimal** numeral system (also called **base 10** or occasionally **denary**) has ten as its base. It is the numerical base most widely used by modern civilizations.

Decimal notation often refers to a base 10 positional notation such as the Hindu-Arabic numeral system or rod calculus; however, it can also be used more generally to refer to non-positional systems such as Roman or Chinese numerals which are also based on powers of ten.

A **decimal number**, or just **decimal**, refers to any number written in decimal notation, although it is more commonly used to refer to numbers that have a fractional part separated from the integer part with a decimal separator (e.g. 11.25).

A decimal may be a terminating decimal, which has a finite fractional part (e.g. 15.600); a repeating decimal, which has an infinite (non-terminating) fractional part made up of a repeating sequence of digits (e.g. 5.123144); or an infinite decimal, which has a fractional part that neither terminates nor has an infinitely repeating pattern (e.g. 3.14159265...). Decimal fractions have terminating decimal representations and other fractions have repeating decimal representations, whereas irrational numbers have infinite non-repeating decimal representations.

Decimal Notation

Decimal notation is the writing of numbers in a base 10 numeral system. Examples are Brahmi numerals, Greek numerals, Hebrew numerals, Roman numerals, and Chinese numerals, as well as the Hindu-Arabic numerals used by speakers of many European languages. Roman numerals have symbols for the decimal powers (1, 10, 100, 1000) and secondary symbols for half these values (5, 50, 500). Brahmi numerals have symbols for the nine numbers 1–9, the nine decades 10–90, plus a symbol for 100 and another for 1000. Chinese numerals have symbols for 1–9, and additional symbols for powers of 10, which in modern usage reach 10^{72}.

However, when people who use Hindu-Arabic numerals speak of decimal notation, they often mean not just decimal numeration, as above, but also decimal fractions, all conveyed as part of a positional system. Positional decimal systems include a zero and use symbols (called digits) for the ten values (0, 1, 2, 3, 4, 5, 6, 7, 8, and 9) to represent any number, no matter how large or how small. These digits are often used with a decimal separator which indicates the start of a fractional part, and with a symbol such as the plus sign + (for positive) or minus sign − (for negative) adjacent to the numeral to indicate whether it is greater or less than zero, respectively.

Positional notation uses positions for each power of ten: units, tens, hundreds, thousands, etc. The position of each digit within a number denotes the multiplier (power of ten) multiplied with that digit—each position has a value ten times that of the position to its right. There were at least two presumably independent sources of positional decimal systems in ancient civilization: the Chinese counting rod system and the Hindu-Arabic numeral system (the latter descended from Brahmi numerals).

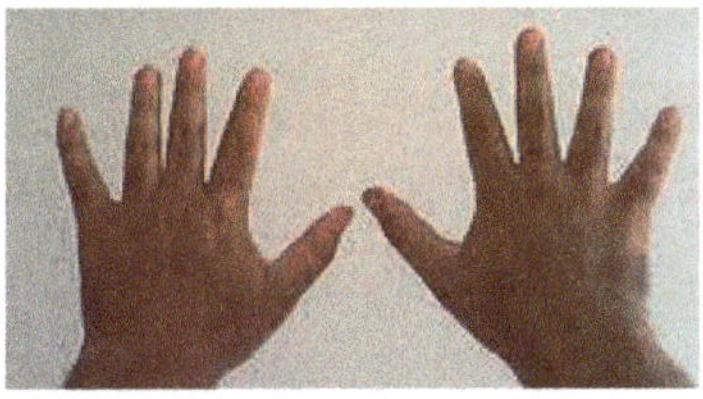

Ten fingers on two hands, the possible starting point of the decimal counting.

Ten is the number which is the count of fingers and thumbs on both hands (or toes on the feet). The English word digit as well as its translation in many languages is also the anatomical term for fingers and toes. In English, decimal (decimus < Lat.) means *tenth*, decimate means *reduce by a tenth*, and denary (denarius < Lat.) means *the unit of ten.*

The symbols for the digits in common use around the globe today are called Arabic numerals by Europeans and Indian numerals by Arabs, the two groups' terms both referring to the culture from which they learned the system. However, the symbols used in different areas are not identical; for instance, Western Arabic numerals (from which the European numerals are derived) differ from the forms used by other Arab cultures.

Decimal Fractions

A **decimal fraction** is a fraction the denominator of which is a power of ten.

Decimal fractions are commonly expressed in decimal notation rather than fraction notation by discarding the denominator and inserting the decimal separator into the numerator at the position from the right corresponding to the power of ten of the denominator and filling the gap with leading zeros if needed, e.g. decimal fractions 8/10, 1489/100, 24/100000, and 58900/10000 are expressed in decimal notation as 0.8, 14.89, 0.00024, 5.8900 respectively. In English-speaking, some Latin American and many Asian countries, a period (**.**) or raised period (·) is used as the decimal separator; in many other countries, particularly in Europe, a comma (**,**) is used.

The integer part, or integral part of a decimal number is the part to the left of the decimal separator. (See also truncation.) The part from the decimal separator to the right is the *fractional part.* It is usual for a decimal number that consists only of a fractional part (mathematically, a *proper fraction*) to have a leading zero in its notation (its *numeral*). This helps disambiguation between a decimal sign and other punctuation, and especially when the negative number sign is indicated, it helps visualize the sign of the numeral as a whole.

Trailing zeros after the decimal point are not necessary, although in science, engineering and statistics they can be retained to indicate a required precision or to show a level of confidence in the accuracy of the number: Although 0.080 and 0.08 are numerically equal, in engineering 0.080 suggests a measurement with an error of up to one part in two thousand (±0.0005), while 0.08 suggests a measurement with an error of up to one in two hundred (see *significant figures*).

Other Rational Numbers

Any rational number with a denominator whose only prime factors are 2 and/or 5 may be precisely expressed as a decimal fraction and has a finite decimal expansion.

$\frac{1}{2} = 0.5$
$\frac{1}{20} = 0.05$
$\frac{1}{5} = 0.2$
$\frac{1}{50} = 0.02$
$\frac{1}{4} = 0.25$
$\frac{1}{40} = 0.025$
$\frac{1}{25} = 0.04$
$\frac{1}{8} = 0.125$
$\frac{1}{125} = 0.008$
$\frac{1}{10} = 0.1$

If a fully reduced rational number's denominator has any prime factors other than 2 or 5, it cannot be expressed as a finite decimal fraction, and has a unique eventually repeating infinite decimal expansion.

$\frac{1}{3} = 0.333333...$ (with 3 repeating)
$\frac{1}{9} = 0.111111...$ (with 1 repeating)

$100 - 1 = 99 = 9 \times 11$:

$\frac{1}{11} = 0.090909...$

$1000 - 1 = 9 \times 111 = 27 \times 37$:

$\frac{1}{27} = 0.037037037...$
$\frac{1}{37} = 0.027027027...$
$\frac{1}{111} = 0.009009009...$

also:

$\frac{1}{81} = 0.012345679012...$ (with 012345679 repeating)

That a rational number must have a finite or recurring decimal expansion can be seen to be a consequence of the long division algorithm, in that there are at most $q - 1$ possible nonzero remainders on division by q, so that the recurring pattern will have a period less than q. For instance, to find 3/7 by long division:

```
  0.4 2 8 5 7 1 4 ...
7)3.0 0 0 0 0 0 0 0
  2 8                     30 ÷ 7 = 4 with a remainder of 2
  2 0
    1 4                   20 ÷ 7 = 2 with a remainder of 6
    6 0
      5 6                 60 ÷ 7 = 8 with a remainder of 4
      4 0
        3 5               40 ÷ 7 = 5 with a remainder of 5
        5 0
          4 9             50 ÷ 7 = 7 with a remainder of 1
          1 0
            7             10 ÷ 7 = 1 with a remainder of 3
            3 0
            2 8           30 ÷ 7 = 4 with a remainder of 2
            2 0
                  etc.
```

The converse to this observation is that every recurring decimal represents a rational number p/q. This is a consequence of the fact that the recurring part of a decimal representation is, in fact, an infinite geometric series which will sum to a rational number. For instance,

$$0.0123123123\cdots = \frac{123}{10000}\sum_{k=0}^{\infty}0.001^k = \frac{123}{10000}\,\frac{1}{1-0.001} = \frac{123}{9990} = \frac{41}{3330}$$

Real Numbers

Every real number has a (possibly infinite) decimal representation; i.e., it can be written as

$$x = \text{sign}\sum_{i\in\mathbb{Z}} a_i\, 10^i$$

where

- sign ∈ {+,−}, which is related to the sign function,
- $\mathbb{Z}$ is the set of all integers (positive, negative, and zero), and
- $a_i \in \{\ 0,1,...,9\ \}$ for all $i \in \mathbb{Z}$ are its decimal digits, equal to zero for all i greater than some number (that number being the common logarithm of $|x|$).

Such a sum converges as more and more negative values of i are included, even if there are infinitely many non-zero a_i.

Rational numbers (e.g., p/q) with prime factors in the denominator other than 2 and 5 (when reduced to simplest terms) have a unique recurring decimal representation.

Non-Uniqueness of Decimal Representation

Consider those rational numbers which have only the factors 2 and 5 in the denominator, i.e., which can be written as $\frac{p}{2^a5^b}$. In this case there is a terminating decimal representation. For instance, $\frac{1}{1} = 1$, $\frac{1}{2} = 0.5$, $\frac{3}{5} = 0.6$, $\frac{3}{25} = 0.12$ and $\frac{1306}{1250} = 1.0448$. Such numbers are the only real numbers which do not have a unique decimal representation, as they can also be written as a representation that has a recurring 9, for instance 1 = 0.99999. $\frac{1}{2} = 0.499999...$, etc. The number $0 = \frac{0}{1}$ is special in that it has no representation with recurring 9.

This leaves the irrational numbers. They also have unique infinite decimal representations, and can be characterised as the numbers whose decimal representations neither terminate nor recur.

So in general the decimal representation is unique, if one excludes representations that end in a recurring 9.

The same trichotomy holds for other base-n positional numeral systems:

- Terminating representation: rational where the denominator divides some n^k
- Recurring representation: other rational
- Non-terminating, non-recurring representation: irrational

A version of this even holds for irrational-base numeration systems, such as golden mean base representation.

Decimal Computation

Decimal computation was carried out in ancient times in many ways, typically in rod calculus, with decimal multiplication table used in ancient China and with sand tables in India and Middle East or with a variety of abaci.

Modern computer hardware and software systems commonly use a binary representation internally (although many early computers, such as the ENIAC or the IBM 650, used decimal representation internally). For external use by computer specialists, this binary representation is sometimes presented in the related octal or hexadecimal systems.

For most purposes, however, binary values are converted to or from the equivalent decimal values for presentation to or input from humans; computer programs express literals in decimal by default. (123.1, for example, is written as such in a computer program, even though many computer languages are unable to encode that number precisely.)

Both computer hardware and software also use internal representations which are effectively decimal for storing decimal values and doing arithmetic. Often this arithmetic is done on data which are encoded using some variant of binary-coded decimal, especially in database implementations, but there are other decimal representations in use (such as in the new IEEE 754 Standard for Floating-Point Arithmetic).

Decimal arithmetic is used in computers so that decimal fractional results can be computed exactly, which is not possible using a binary fractional representation. This is often important for financial and other calculations.

1/2	1	2	3	4	5	6	7	8	9	10	20	30	40	50	60	70	80	90	
45	90	180	270	360	450	540	630	720	810	900	1800	2700	3600	4500	5400	6300	7200	8100	90
40	80	160	240	320	400	480	560	640	720	800	1600	2400	3200	4000	4800	5600	6400	7200	80
35	70	140	210	280	350	420	490	560	630	700	1400	2100	2800	3500	4200	4900	5600	6300	70
30	60	120	180	240	300	360	420	480	540	600	1200	1800	2400	3000	3600	4200	4800	5400	60
25	50	100	150	200	250	300	350	400	450	500	1000	1500	2000	2500	3000	3500	4000	4500	50
20	40	80	120	160	200	240	280	320	360	400	800	1200	1600	2000	2400	2800	3200	3600	40
15	30	60	90	120	150	180	210	240	270	300	600	900	1200	1500	1800	2100	2400	2700	30
10	20	40	60	80	100	120	140	160	180	200	400	600	800	1000	1200	1400	1600	1800	20
5	10	20	30	40	50	60	70	80	90	100	200	300	400	500	600	700	800	900	10
4.5	9	18	27	36	45	54	63	72	81	90	180	270	360	450	540	630	720	810	9
4	8	16	24	32	40	48	56	64	72	80	160	240	320	400	480	560	640	720	8
3.5	7	14	21	28	35	42	49	56	63	70	140	210	280	350	420	490	560	630	7
3	6	12	18	24	30	36	42	48	54	60	120	180	240	300	360	420	480	540	6
2.5	5	10	15	20	25	30	35	40	45	50	100	150	200	250	300	350	400	450	5
2	4	8	12	16	20	24	28	32	36	40	80	120	160	200	240	280	320	360	4
1.5	3	6	9	12	15	18	21	24	27	30	60	90	120	150	180	210	240	270	3
1	2	4	6	8	10	12	14	16	18	20	40	60	80	100	120	140	160	180	2
0.5	1	2	3	4	5	6	7	8	9	10	20	30	40	50	60	70	80	90	1
0.25	0.5	1	1.5	2	2.5	3	3.5	4	4.5	5	10	15	20	25	30	35	40	45	1/2

Diagram of the world's earliest decimal multiplication table (c. 305 BC) from the Warring States period

Many ancient cultures calculated with numerals based on ten, as human hands typically have 10 digits. Egyptian hieroglyphs, in evidence since around 3000 BC, used a purely decimal system, just as the Cretan hieroglyphs (ca. 1625−1500 BC) of the Minoans whose numerals are closely based on the Egyptian model. The decimal system was handed down to the consecutive Bronze Age cultures of Greece, including Linear A (ca. 18th century BC−1450 BC) and Linear B (ca. 1375−1200 BC) — the number system of classical Greece also used powers of ten, including, like the Roman numerals did, an intermediate base of 5. Notably, the polymath Archimedes (ca. 287–212 BC) invented a decimal positional system in his Sand Reckoner which was based on 10^8 and later led the German mathematician Carl Friedrich Gauss to lament what heights science would have already reached in his days if Archimedes had fully realized the potential of his ingenious discovery. The Hittites hieroglyphs (since 15th century BC), just like the Egyptian and early numerals in Greece, was strictly decimal.

The Egyptian hieratic numerals, the Greek alphabet numerals, the Hebrew alphabet numerals, the Roman numerals, the Chinese numerals and early Indian Brahmi numerals are all non-positional decimal systems, and required large numbers of symbols. For instance, Egyptian numerals used different symbols for 10, 20, to 90, 100, 200, to 900, 1000, 2000, 3000, 4000, to 10,000. The world's earliest positional decimal system was the Chinese rod calculus

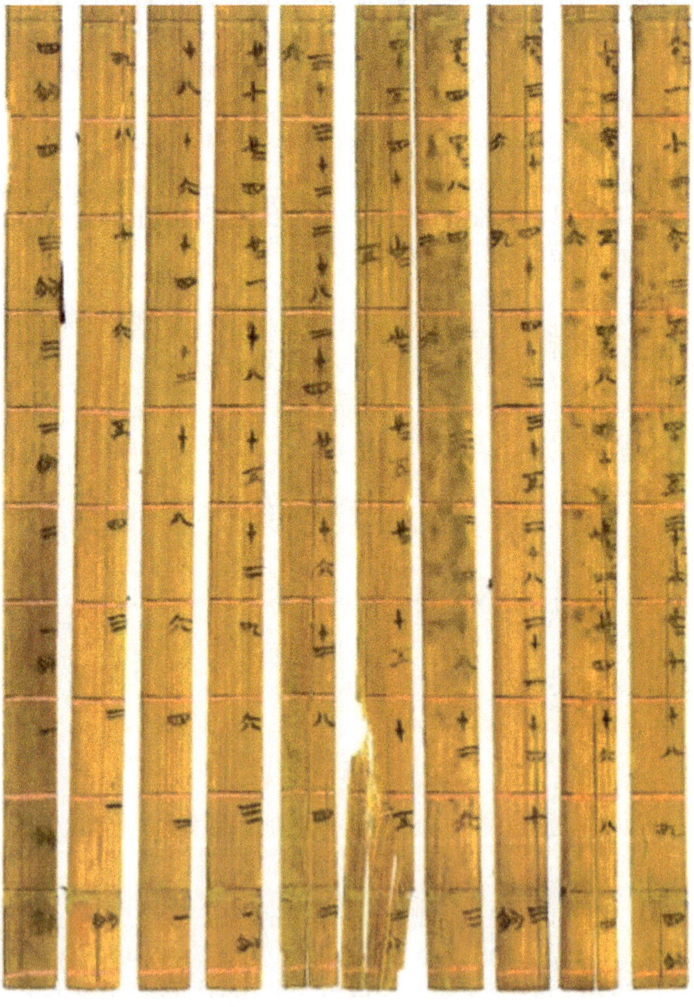

The world's earliest decimal multiplication table was made from bamboo slips, dating from 305 BC, during the Warring States period in China.

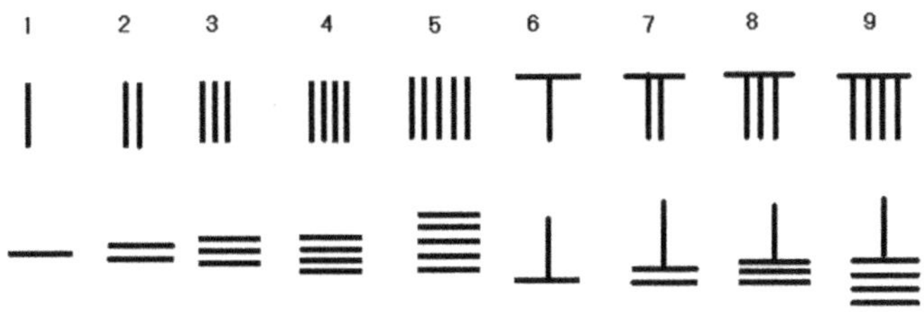

The world's earliest positional decimal system Upper row vertical form Lower row horizontal form

History of Decimal Fractions

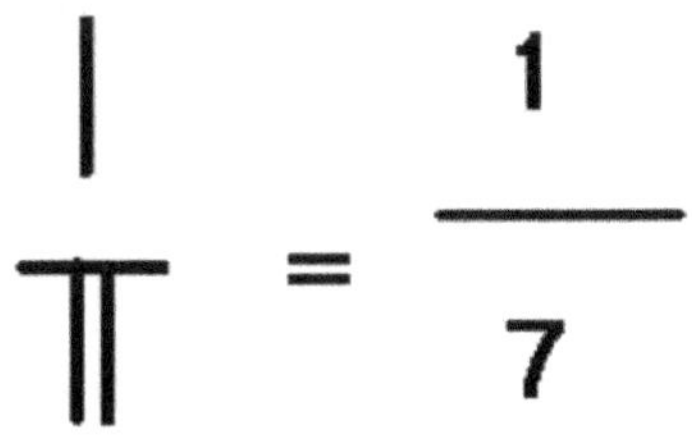

counting rod decimal fraction 1/7

According to Joseph Needham and Lam Lay Yong, decimal fractions were first developed and used by the Chinese in the 1st century BC, and then spread to the Middle East and from there to Europe. The written Chinese decimal fractions were non-positional. However, counting rod fractions were positional.

Qin Jiushao in his book Mathematical Treatise in Nine Sections (1247) denoted 0.96644 by

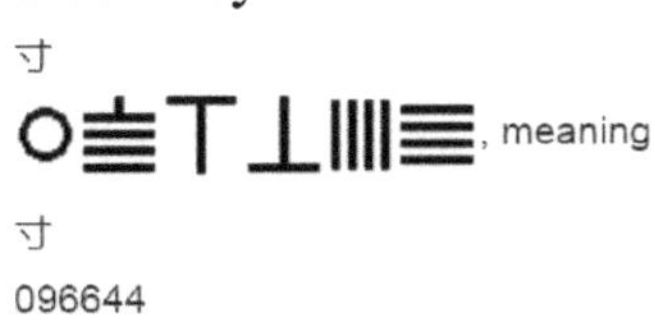

J. Lennart Berggren notes that positional decimal fractions appear for the first time in a book by the Arab mathematician Abu'l-Hasan al-Uqlidisi written in the 10th century. The Jewish mathematician Immanuel Bonfils used decimal fractions around 1350, anticipating Simon Stevin, but did not develop any notation to represent them. The Persian mathematician Jamshīd al-Kāshī claimed to have discovered decimal fractions himself in the 15th century. Al Khwarizmi introduced fraction to Islamic countries in the early 9th century, his fraction presentation was an exact copy of traditional Chinese mathematical fraction from The Mathematical Classic of Sunzi. This form of fraction with numerator on top and denominator at bottom without a horizontal bar was also used by 10th century Abu'l-Hasan al-Uqlidisi and 15th century Jamshīd al-Kāshī's work "Arithmetic Key".

Number: 184.54290

Simon Stevin's notation: 184⓪5①4②2③9④0

A forerunner of modern European decimal notation was introduced by Simon Stevin in the 16th century.

Natural Languages

The ingenious method of expressing every possible number using a set of ten symbols emerged in India. Several Indian languages show a straightforward decimal system. Many Indo-Aryan and Dravidian languages have numbers between 10 and 20 expressed in a regular pattern of addition to 10.

The Hungarian language also uses a straightforward decimal system. All numbers between 10 and 20 are formed regularly (e.g. 11 is expressed as "tizenegy" literally "one on ten"), as with those between 20 and 100 (23 as "huszonhárom" = "three on twenty").

A straightforward decimal rank system with a word for each order (10 十, 100 百, 1000 千, 10,000 万), and in which 11 is expressed as *ten-one* and 23 as *two-ten-three*, and 89,345 is expressed as 8 (ten thousands) 万 9 (thousand) 千 3 (hundred) 百 4 (tens) 十 5 is found in Chinese, and in Vietnamese with a few irregularities. Japanese, Korean, and Thai have imported the Chinese decimal system. Many other languages with a decimal system have special words for the numbers between 10 and 20, and decades. For example, in English 11 is "eleven" not "ten-one" or "one-teen".

Incan languages such as Quechua and Aymara have an almost straightforward decimal system, in which 11 is expressed as *ten with one* and 23 as *two-ten with three*.

Some psychologists suggest irregularities of the English names of numerals may hinder children's counting ability.

Other Bases

Some cultures do, or did, use other bases of numbers.

- Pre-Columbian Mesoamerican cultures such as the Maya used a base-20 system (presumably using all twenty fingers and toes).
- The Yuki language in California and the Pamean languages in Mexico have octal (base-8) systems because the speakers count using the spaces between their fingers rather than the fingers themselves.
- The existence of a non-decimal base in the earliest traces of the Germanic languages, is attested by the presence of words and glosses meaning that the count is in decimal (cognates to ten-count or tenty-wise), such would be expected if normal counting is not decimal, and unusual if it were. Where this counting system is known, it is based on the long hundred of 120 in number, and a long thousand of 1200 in number. The descriptions like 'long' only appear after the small hundred of 100 in number appeared with the Christians. Gordon's Introduction to Old Norse p 293, gives number names that belong to this system. An expression cognate to 'one hundred and eighty' is translated to 200, and the cognate to 'two hundred' is translated at 240. Goodare details the use of the long hundred in Scotland in the Middle Ages, giving examples, calculations where the carry implies i C (i.e. one hundred) as 120, etc. That

the general population were not alarmed to encounter such numbers suggests common enough use. It is also possible to avoid hundred-like numbers by using intermediate units, such as stones and pounds, rather than a long count of pounds. Goodare gives examples of numbers like vii score, where one avoids the hundred by using extended scores. There is also a paper by W.H. Stevenson, on ‹Long Hundred and its uses in England'.

- Many or all of the Chumashan languages originally used a base-4 counting system, in which the names for numbers were structured according to multiples of 4 and 16.
- Many languages use quinary (base-5) number systems, including Gumatj, Nunggubuyu, Kuurn Kopan Noot and Saraveca. Of these, Gumatj is the only true 5–25 language known, in which 25 is the higher group of 5.
- Some Nigerians use duodecimal systems. So did some small communities in India and Nepal, as indicated by their languages.
- The Huli language of Papua New Guinea is reported to have base-15 numbers. *Ngui* means 15, *ngui ki* means 15 × 2 = 30, and *ngui ngui* means 15 × 15 = 225.
- Umbu-Ungu, also known as Kakoli, is reported to have base-24 numbers. *Tokapu* means 24, *tokapu talu* means 24 × 2 = 48, and *tokapu tokapu* means 24 × 24 = 576.
- Ngiti is reported to have a base-32 number system with base-4 cycles.
- The Ndom language of Papua New Guinea is reported to have base-6 numerals. *Mer* means 6, *mer an thef* means 6 × 2 = 12, *nif* means 36, and *nif thef* means 36×2 = 72.

PERCENTAGE

In mathematics, a **percentage** is a number or ratio expressed as a fraction of 100. It is often denoted using the percent sign, "%", or the abbreviations "pct.", "pct"; sometimes the abbreviation "pc" is also used. A percentage is a dimensionless number (pure number).

Examples

For example, 45% (read as "forty-five percent") is equal to $^{45}/_{100}$, or 0.45. Percentages which are not ratios are expressed as a relationship of the part to the whole. For example, if the whole was the total number of students in a class, when 50% of the students in the class were male that means that 50 out of every 100 students were male. If there were 1000 students then 500 would be male. A related system which expresses a number as a fraction of 1,000 uses the terms "per mil" and "millage". Percentages are used as ratios to express how large or small one quantity is relative to another quantity. The first quantity usually represents a part of, or a change in, the second quantity. For example, an increase of $0.15 on a price of $2.50 is an increase by a fraction of 0.15/2.50 = 0.06. Expressed as a percentage, this is therefore a 6% increase. While percentage values may often be limited to lie between 0 and 100, there is no mathematical restriction and percentages may take on other values. For example, it is common to refer to 111% or −35%, especially for percent changes and comparisons.

In Ancient Rome, long before the existence of the decimal system, computations were often made in fractions which were multiples of $^{1}/_{100}$. For example, Augustus levied a tax of $^{1}/_{100}$ on goods sold at auction known as *centesima rerum venalium*. Computation with these fractions was similar to computing percentages. As denominations of money grew in the Middle Ages, computations with a denominator of 100 became more standard and from the late 15th century to the early 16th century it became common for arithmetic texts to include such computations. Many of these texts applied these methods to profit and loss, interest rates, and the Rule of Three. By the 17th century it was standard to quote interest rates in hundredths.

Percent Sign

The term "per cent" is derived from the Latin *per centum*, meaning "by the hundred". The sign for "per cent" evolved by gradual contraction of the Italian term *per cento*, meaning "for a hundred". The "per" was often abbreviated as "p." and eventually disappeared entirely. The "cento" was contracted to two circles separated by a horizontal line, from which the modern "%" symbol is derived.

%

A percent sign

Calculations

The percent value is computed by multiplying the numeric value of the ratio by 100. For example, to find 50 apples as a percentage of 1250 apples, first compute the ratio $^{50}/_{1250} = 0.04$, and then multiply by 100 to obtain 4%. The percent value can also be found by multiplying first, so in this example the 50 would be multiplied by 100 to give 5,000, and this result would be divided by 1250 to give 4%.

To calculate a percentage of a percentage, convert both percentages to fractions of 100, or to decimals, and multiply them. For example, 50% of 40% is:

$^{50}/_{100} \times {}^{40}/_{100} = 0.50 \times 0.40 = 0.20 = {}^{20}/_{100} = 20\%$. It is not correct to divide by 100 and use the percent sign at the same time. (E.g. $25\% = {}^{25}/_{100} = 0.25$, not $^{25\%}/_{100}$, which actually is $^{25}/_{100}/100 = 0.0025$. A term such as $^{100}/_{100}\%$ would also be incorrect, this would be read as 1 percent even if the intent was to say 100%.)

Whenever we talk about a percentage, it is important to specify what it is relative to, i.e. what is the total that corresponds to 100%. The following problem illustrates this point.

In a certain college 60% of all students are female, and 10% of all students are computer science majors. If 5% of female students are computer science majors, what percentage of computer science majors are female?

We are asked to compute the ratio of female computer science majors to all computer science majors. We know that 60% of all students are female, and among these 5% are computer science majors, so we conclude that $^{60}/_{100} \times {}^{5}/_{100} = {}^{3}/_{100}$ or 3% of all students are female computer science majors. Dividing this by the 10% of all students that are computer science majors, we arrive at the answer: $^{3\%}/_{10\%} = {}^{30}/_{100}$ or 30% of all computer science majors are female.

This example is closely related to the concept of conditional probability.

Percentage Increase and Decrease

Sometimes due to inconsistent usage, it is not always clear from the context what a percentage is relative to. When speaking of a "10% rise" or a "10% fall" in a quantity, the usual interpretation is that this is relative to the *initial value* of that quantity. For example, if an item is initially priced at $200 and the price rises 10% (an increase of $20), the new price will be $220. Note that this final price is 110% of the initial price (100% + 10% = 110%).

Some other examples of percent changes:

- An increase of 100% in a quantity means that the final amount is 200% of the initial amount (100% of initial + 100% of increase = 200% of initial); in other words, the quantity has doubled.
- An increase of 800% means the final amount is 9 times the original (100% + 800% = 900% = 9 times as large).
- A decrease of 60% means the final amount is 40% of the original (100% – 60% = 40%).
- A decrease of 100% means the final amount is *zero* (100% – 100% = 0%).

In general, a change of x percent in a quantity results in a final amount that is 100 + x percent of the original amount (equivalently, $1 + 0.01x$ times the original amount).

Compounding Percentages

It is important to understand that percent changes, as they have been discussed here, *do not add* in the usual way, if applied sequentially. For example, if the 10% increase in price considered earlier (on the $200 item, raising its price to $220) is followed by a 10% decrease in the price (a decrease of $22), the final price will be $198, *not* the original price of $200. The reason for the apparent discrepancy is that the two percent changes (+10% and −10%) are measured relative to *different* quantities ($200 and $220, respectively), and thus do not "cancel out".

In general, if an increase of x percent is followed by a decrease of x percent, and the initial amount was p, the final amount is $p(1 + 0.01x)(1 - 0.01x) = p(1 - (0.01x)^2)$; thus the net change is an overall decrease by x percent *of* x percent (the square of the original percent change when expressed as a decimal number). Thus, in the above example, after an increase and decrease of $x = 10$ percent, the final amount, $198, was 10% of 10%, or 1%, less than the initial amount of $200. The net change is the same for a decrease of x percent followed by an increase of x percent; the final amount is $p(1 + 0.01x)(1 - 0.01x) = p(1 - (0.01x)^2)$.

This can be expanded for a case where you do not have the same percent change. If the initial percent change is x and the second percent change is y, and the initial amount was p, then the final amount is $p(1 + 0.01x)(1 + 0.01y)$. To change the above example, after an increase of $x = 10$ percent and decrease of $y = -5$ percent, the final amount, \$209, is 4.5% more than the initial amount of \$200.

As shown above, percent changes can be applied in any order and have the same effect.

In the case of interest rates, a very common but mathematically incorrect way to state that an interest rate rose from 10% to 15%, for example, is to say that the interest rate increased by 5 percent, when in fact what is meant is that the interest rate increased by 5 percentage points (pp). As a result, people often don't understand until later the shocking truth that their monthly payments on a loan will increase by 20 percent when banks or journalists (incorrectly) talk about a 2% increase when they actually mean an increase from 10% to 12%. The same confusion between the different concepts of percent(age) and percentage points causes major misunderstandings when journalists regularly incorrectly report about election results, for example. In talking about bonds, it is common to refer to an increase of one percentage point as an increase of *100 basis points*.

Word and Symbol

In British English, *percent* is sometimes written as two words (*per cent*, although *percentage* and *percentile* are written as one word). In American English, *percent* is the most common variant (but cf. *per mille* written as two words).

In the early part of the twentieth century, there was a dotted abbreviation form *"per cent."*, as opposed to *«per cent"*. The form "per cent." is still in use as a part of the highly formal language found in certain documents like commercial loan agreements (particularly those subject to, or inspired by, common law), as well as in the Hansard transcripts of British Parliamentary proceedings. The term has been attributed to Latin *per centum*. The concept of considering values as parts of a hundred is originally Greek. The symbol for percent (%) evolved from a symbol abbreviating the Italian *per cento*. In some other languages, the form *procent* or *prosent* is used instead. Some languages use both a word derived from *percent* and an expression in that language meaning the same thing, e.g. Romanian *procent* and *la sută* (thus, *10%* can be read or sometimes written *ten for [each] hundred*, similarly with the English *one out of ten*). Other abbreviations are rarer, but sometimes seen.

Grammar and style guides often differ as to how percentages are to be written. For instance, it is commonly suggested that the word percent (or per cent) be spelled out in all texts, as in "1 percent" and not "1%". Other guides prefer the word to be written out in humanistic texts, but the symbol to be used in scientific texts. Most guides agree that they always be written with a numeral, as in "5 percent" and not "five percent", the only exception being at the beginning of a sentence: "Ten percent of all writers love style guides." Decimals are also to be used instead of fractions, as in "3.5 percent of the gain" and not "3 1/2 percent of the gain". It is also widely accepted to use the percent symbol (%) in tabular and graphic material.

In line with common English practice, style guides—such as *The Chicago Manual of Style*—generally state that the number and percent sign are written without any space in between. However, the International System of Units and the ISO 31-0 standard require a space.

Other Uses

The word "percentage" is often a misnomer in the context of sports statistics, when the referenced number is expressed as a decimal proportion, not a percentage: "The Phoenix Suns' Shaquille O'Neal led the NBA with a .609 field goal percentage (FG%) during the 2008–09 season." (O'Neal made 60.9% of his shots, not 0.609%.) Likewise, the winning percentage of a team, the fraction of matches that the club has won, is also usually expressed as a decimal proportion; a team that has a .500 winning percentage has won 50% of their matches. The practice is probably related to the similar way that batting averages are quoted.

As "percent" it is used to describe the steepness of the slope of a road or railway, formula for which is 100 × rise/run which could also be expressed as the tangent of the angle of inclination times 100. This is the ratio of distances a vehicle would advance vertically and horizontally, respectively, when going up- or downhill, expressed in percent.

Percentage is also used to express composition of a mixture by mass percent and mole percent.

Related Units

- Percentage point
- Per mille (‰) 1 part in 1,000
- Basis point (‱) 1 part in 10,000
- Per cent mille (pcm) 1 part in 100,000
- Parts-per notation
- Grade (slope)
- Per-unit system

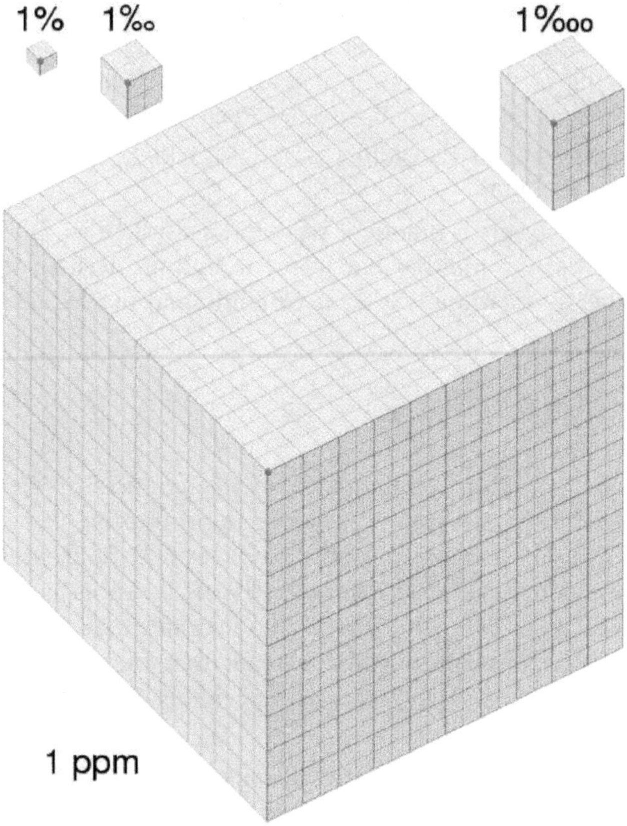

Visualisation of 1%, 1‰, 1‱ and 1 ppm as the fraction of the red cube to its respective block (click for larger version)

REFERENCES

1. http://www.telegraph.co.uk/finance/economics/11329769/Eurozone-officially-falls-into-deflation-piling-pressure-on-ECB.html
2. Bennett, Jeffrey; Briggs, William (2005), Using and Understanding Mathematics / A Quantitative Reasoning Approach (3rd ed.), Pearson Addison Wesley, p. 134, ISBN 0-321-22773-5
3. Smith, D.E. (1958) [1951]. History of Mathematics. 2. Courier Dover Publications. pp. 247–249. ISBN 0-486-20430-8.
4. American Heritage Dictionary of the English Language, 3rd ed. (1992) Houghton Mifflin
5. Smith p. 250
6. Brians, Paul. "Percent/per cent". Common Errors in English Usage. Washington State University. Retrieved 22 November 2010.
7. "Percent (per cent)". Oxford Dictionaries. Retrieved 22 November 2010.

8. "Percent". Oxford English Dictionary (3rd ed.). Oxford University Press. September 2005. (Subscription or UK public library membership required.)
9. "The Chicago Manual of Style". University of Chicago Press. 2003. Retrieved 2007-01-05.
10. "The International System of Units" (PDF). International Bureau of Weights and Measures. 2006. Retrieved 2007-08-06.
11. "ISO 31-0 — Quantities and units – Part 0: General principles". International Organization for Standardization. 1999-12-22. Retrieved 2007-01-05.
12. The History of Arithmetic, Louis Charles Karpinski, 200pp, Rand McNally & Company, 1925.
13. Lam Lay Yong & Ang Tian Se (2004) Fleeting Footsteps. Tracing the Conception of Arithmetic and Algebra in Ancient China, Revised Edition, World Scientific, Singapore.
14. The overline in 5.123144 indicates that the '144' sequence repeats itself indefinitely, i.e. 5.123144144144144....
15. "Decimal Fraction". Encyclopedia of Mathematics. Retrieved 2013-06-18.
16. Math Made Nice-n-Easy. Piscataway, N.J.: Research Education Association. 1999. p. 141. ISBN 0-87891-200-2.
17. Fingers or Fists? (The Choice of Decimal or Binary Representation), Werner Buchholz, Communications of the ACM, Vol. 2 #12, pp3–11, ACM Press, December 1959.
18. Schmid, Hermann (1983) [1974]. Decimal Computation (1 (reprint) ed.). Malabar, Florida, USA: Robert E. Krieger Publishing Company. ISBN 0-89874-318-4.
19. Schmid, Hermann (1974). Decimal Computation (1 ed.). Binghamton, New York, USA: John Wiley & Sons. ISBN 0-471-76180-X.
20. Decimal Floating-Point: Algorism for Computers, Cowlishaw, M. F., Proceedings 16th IEEE Symposium on Computer Arithmetic, ISBN 0-7695-1894-X, pp104-111, IEEE Comp. Soc., June 2003
21. Decimal Arithmetic - FAQ
22. Decimal Floating-Point: Algorism for Computers, Cowlishaw, M. F., Proceedings 16th IEEE Symposium on Computer Arithmetic (ARITH 16), ISBN 0-7695-1894-X, pp. 104–111, IEEE Comp. Soc., June 2003
23. Egyptian numerals
24. Georges Ifrah: From One to Zero. A Universal History of Numbers, Penguin Books, 1988, ISBN 0-14-009919-0, pp. 200-213 (Egyptian Numerals)
25. Graham Flegg: Numbers: their history and meaning, Courier Dover Publications, 2002, ISBN 978-0-486-42165-0, p. 50
26. Georges Ifrah: From One to Zero. A Universal History of Numbers, Penguin Books, 1988, ISBN 0-14-009919-0, pp.213-218 (Cretan numerals)
27. Greek numerals

28. Menninger, Karl: Zahlwort und Ziffer. Eine Kulturgeschichte der Zahl, Vandenhoeck und Ruprecht, 3rd. ed., 1979, ISBN 3-525-40725-4, pp. 150-153
29. Georges Ifrah: From One to Zero. A Universal History of Numbers, Penguin Books, 1988, ISBN 0-14-009919-0, pp. 218f. (The Hittite hieroglyphic system)
30. Lam Lay Yong et al The Fleeting Footsteps p 137-139
31. Lam Lay Yong, "The Development of Hindu-Arabic and Traditional Chinese Arithmetic", Chinese Science, 1996 p38, Kurt Vogel notation
32. Joseph Needham (1959). "Decimal System". Science and Civilisation in China, Volume III, Mathematics and the Sciences of the Heavens and the Earth. Cambridge University Press.
33. Jean-Claude Martzloff, A History of Chinese Mathematics, Springer 1997 ISBN 3-540-33782-2
34. Berggren, J. Lennart (2007). "Mathematics in Medieval Islam". In Katz, Victor J. The Mathematics of Egypt, Mesopotamia, China, India, and Islam: A Sourcebook. Princeton University Press. p. 530. ISBN 978-0-691-11485-9.
35. Gandz, S.: The invention of the decimal fractions and the application of the exponential calculus by Immanuel Bonfils of Tarascon (c. 1350), Isis 25 (1936), 16–45.
36. Lam Lay Yong, "A Chinese Genesis, Rewriting the history of our numeral system", Archive for History of Exact Science 38: 101–108.
37. B. L. van der Waerden (1985). A History of Algebra. From Khwarizmi to Emmy Noether. Berlin: Springer-Verlag.
38. "Indian numerals". Ancient Indian mathematics. Retrieved 2015-05-22.
39. Azar, Beth (1999). "English words may hinder math skills development". American Psychology Association Monitor. 30 (4). Archived from the original on 2007-10-21.
40. Avelino, Heriberto (2006). "The typology of Pame number systems and the limits of Mesoamerica as a linguistic area"(PDF). Linguistic Typology. 10 (1): 41–60. doi:10.1515/LINGTY.2006.002.
41. Marcia Ascher. "Ethnomathematics: A Multicultural View of Mathematical Ideas". The College Mathematics Journal. Retrieved 2007-04-13.
42. McClean, R. J. (July 1958), "Observations on the Germanic numerals", German Life and Letters, 11 (4): 293–299, doi:10.1111/j.1468-0483.1958.tb00018.x, Some of the Germanic languages appear to show traces of an ancient blending of the decimal with the vigesimal system.
43. Voyles, Joseph (October 1987), "The cardinal numerals in pre-and proto-Germanic", The Journal of English and Germanic Philology, 86 (4): 487–495, JSTOR 27709904.
44. There is a surviving list of Ventureño language number words up to 32 written down by a Spanish priest ca. 1819. "Chumashan Numerals" by Madison S. Beeler, in Native American Mathematics, edited by Michael P. Closs (1986), ISBN 0-292-75531-7.

45. Hammarström, Harald (17 May 2007). "Rarities in Numeral Systems". In Wohlgemuth, Jan; Cysouw, Michael. Rethinking Universals: How rarities affect linguistic theory(PDF). Empirical Approaches to Language Typology. 45. Berlin: Mouton de Gruyter (published 2010). Archived from the original (PDF) on 19 August 2007.

46. Harris, John (1982). Hargrave, Susanne, ed. Facts and fallacies of aboriginal number systems (PDF). Work Papers of SIL-AAB Series B. 8. pp. 153–181.

47. Dawson, J. "Australian Aborigines: The Languages and Customs of Several Tribes of Aborigines in the Western District of Victoria (1881), p. xcviii.

48. Matsushita, Shuji (1998). Decimal vs. Duodecimal: An interaction between two systems of numeration. 2nd Meeting of the AFLANG, October 1998, Tokyo. Archived from the original on 2008-10-05. Retrieved 2011-05-29.

49. Mazaudon, Martine (2002). "Les principes de construction du nombre dans les langues tibéto-birmanes". In François, Jacques. La Pluralité (PDF). Leuven: Peeters. pp. 91–119. ISBN 90-429-1295-2

50. Cheetham, Brian (1978). "Counting and Number in Huli". Papua New Guinea Journal of Education. 14: 16–35. Archived from the original on 2007-09-28.

51. Bowers, Nancy; Lepi, Pundia (1975). "Kaugel Valley systems of reckoning" (PDF). Journal of the Polynesian Society. 84 (3): 309–324.

52. Owens, Kay (2001), "The Work of Glendon Lean on the Counting Systems of Papua New Guinea and Oceania", Mathematics Education Research Journal, 13 (1): 47–71, doi:10.1007/BF03217098

53. H. Wu, "The Mis-Education of Mathematics Teachers", Notices of the American Mathematical Society, Volume 58, Issue 03 (March 2011), page 374

54. Ambrose, Gavin; et al. (2006). The Fundamentals of Typography, (2nd ed.). Lausanne: AVA Publishing. p. 74. ISBN 978-2-940411-76-4..

55. Bringhurst, Robert (2002). "5.2.5: Use the Virgule with Words and Dates, the Solidus with Split-level Fractions". The Elements of Typographic Style (3rd ed.). Point Roberts: Hartley & Marks. pp. 81–82. ISBN 978-0-88179-206-5.

56. "virgule, n.". Oxford English Dictionary (1st ed.). Oxford: Oxford University Press. 1917.

57. "solidus, n.1". Oxford English Dictionary (1st ed.). Oxford: Oxford University Press. 1913..

58. Weisstein, Eric W. "Common Fraction". MathWorld.

59. "World Wide Words: Vulgar fractions". World Wide Words. Retrieved 2014-10-30.

60. Weisstein, Eric W. "Improper Fraction". MathWorld.

61. Laurel (31 March 2004). "Math Forum – Ask Dr. Math:Can Negative Fractions Also Be Proper or Improper?". Retrieved 2014-10-30.

62. "New England Compact Math Resources".

63. Greer, A. (1986). New comprehensive mathematics for 'O' level (2nd ed., reprinted. ed.). Cheltenham: Thornes. p. 5. ISBN 9780859501590. Retrieved 2014-07-29.
64. Trotter, James (1853). A complete system of arithmetic. p. 65.
65. Barlow, Peter (1814). A new mathematical and philosophical dictionary.
66. "Fraction". Encyclopedia of Mathematics. 2012-04-06. Retrieved 2012-08-15.
67. Galen, Leslie Blackwell (March 2004). "Putting Fractions in Their Place" (PDF). American Mathematical Monthly. 111 (3).
68. "built fraction". allbusiness.com glossary. Retrieved 2013-06-18.
69. "piece fraction". allbusiness.com glossary. Retrieved 2013-06-18.
70. Eves, Howard (1990). An introduction to the history of mathematics (6th ed.). Philadelphia: Saunders College Pub. ISBN 0-03-029558-0.
71. Milo Gardner (December 19, 2005). "Math History". Retrieved 2006-01-18. See for examples and an explanation.
72. Miller, Jeff (22 December 2014). "Earliest Uses of Various Mathematical Symbols". Retrieved 15 February 2016..
73. Filliozat (2004), p. 152
74. Cajori, Florian (1928). A History of Mathematical Notations (PDF). 1. La Salle, Illinois: Open Court Publishing Company. p. 269.
75. Cajori (1928), pg.89
76. A Source Book in Mathematics 1200–1800. New Jersey: Princeton University Press. 1986. ISBN 0-691-02397-2.
77. Die Rechenkunst bei Ǧamšīd b. Mas'ūd al-Kāšī. Wiesbaden: Steiner. 1951.
78. Berggren, J. Lennart (2007). "Mathematics in Medieval Islam". The Mathematics of Egypt, Mesopotamia, China, India, and Islam: A Sourcebook. Princeton University Press. p. 518. ISBN 978-0-691-11485-9.
79. "MacTutor's al-Uqlidisi biography". Retrieved 2011-11-22.
80. "Common Core State Standards for Mathematics" (PDF). Common Core State Standards Initiative. 2010. p. 85. Retrieved 2013-10-10.

CHAPTER 3

Quadratic Equation

Chapter Contents

A quadratic equation consists of a single variable with three terms in the standard form: $ax^2 + bx + c = 0$. The first quadratic equations were developed as a method used by Babylonian mathematicians around 2000 BC to solve simultaneous equations. Quadratic equations can be applied to problems in physics involving parabolic motion, path, shape, and stability. Several methods have evolved to simplify the solution of such equations for the variable x. Any number of quadratic equation solvers, in which the values of the quadratic equation coefficients can be entered and automatically calculated, can be found online.

INTRODUCTION

In elementary algebra, a **quadratic equation** (from the Latin quadratus for "square") is any equation having the form

$$ax^2 + bx + c = 0$$

where x represents an unknown, and a, b, and c represent known numbers such that a is not equal to 0. If a = 0, then the equation is linear, not quadratic. The numbers a, b, and c are the coefficients of the equation, and may be distinguished by calling them, respectively, the quadratic coefficient, the linear coefficient and the constant or free term.

Because the quadratic equation involves only one unknown, it is called "univariate". The quadratic equation only contains powers of x that are non-negative integers, and therefore it is a polynomial equation, and in particular it is a second degree polynomial equation since the greatest power is two.

Quadratic equations can be solved by a process known in American English as factoring and in other varieties of English as factorising, by completing the square, by using the quadratic formula, or by graphing. Solutions to problems equivalent to the quadratic equation were known as early as 2000 BC.

$$x = \frac{-b \pm \sqrt{b^2 - 4ac}}{2a}$$

The quadratic formula for the roots of the general quadratic equation

SOLVING THE QUADRATIC EQUATION

A quadratic equation with real or complex coefficients has two solutions, called roots. These two solutions may or may not be distinct, and they may or may not be real.

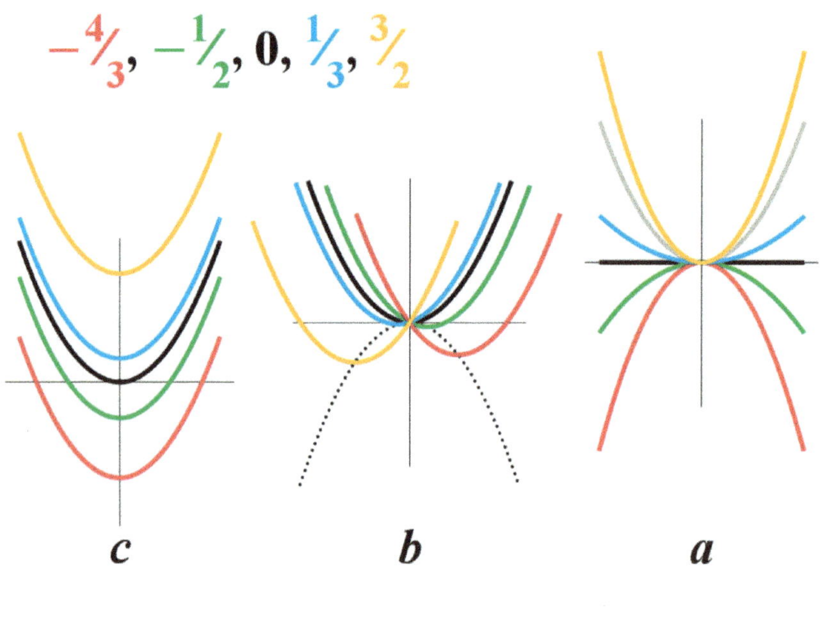

Figure 1. Plots of quadratic function y = ax² + bx + c, varying each coefficient separately while the other coefficients are fixed (at values a = 1, b = 0, c = 0)

Factoring by Inspection

It may be possible to express a quadratic equation $ax^2 + bx + c = 0$ as a product $(px + q)(rx + s) = 0$. In some cases, it is possible, by simple inspection, to determine values of p, q, r, and s that make the two forms equivalent to one another. If the quadratic equation is written in the second form, then the "Zero Factor Property" states that the quadratic equation is satisfied if $px + q = 0$ or $rx + s = 0$. Solving these two linear equations provides the roots of the quadratic.

For most students, factoring by inspection is the first method of solving quadratic equations to which they are exposed. If one is given a quadratic equation in the form $x^2 + bx + c = 0$, the sought factorization has the form $(x + q)(x + s)$, and one has to find two numbers q and s that add up to b and whose product is c (this is sometimes called "Vieta's rule" and is related to Vieta's formulas). As an example, $x^2 + 5x + 6$ factors as $(x + 3)(x + 2)$. The more general case where a does not equal 1 can require a considerable effort in trial and error guess-and-check, assuming that it can be factored at all by inspection.

Except for special cases such as where $b = 0$ or $c = 0$, factoring by inspection only works for quadratic equations that have rational roots. This means that the great majority of quadratic equations that arise in practical applications cannot be solved by factoring by inspection.

Completing the Square

The process of completing the square makes use of the algebraic identity

$$x^2 + 2hx + h^2 = (x + h)^2,$$

which represents a well-defined algorithm that can be used to solve any quadratic equation. Starting with a quadratic equation in standard form, $ax^2 + bx + c = 0$

- Divide each side by a, the coefficient of the squared term.
- Subtract the constant term c/a from both sides.
- Add the square of one-half of b/a, the coefficient of x, to both sides. This "completes the square", converting the left side into a perfect square.
- Write the left side as a square and simplify the right side if necessary.
- Produce two linear equations by equating the square root of the left side with the positive and negative square roots of the right side.
- Solve the two linear equations.

We illustrate use of this algorithm by solving $2x^2 + 4x - 4 = 0$

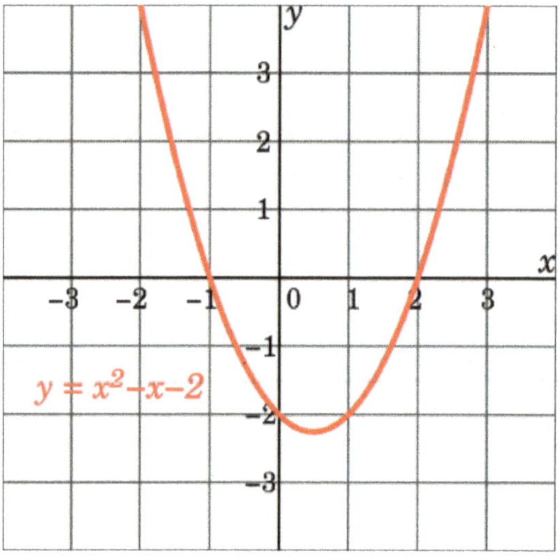

Figure 2. For the quadratic functiony = $x^2 - x - 2$, the points where the graph crosses the x-axis, x = −1 and x = 2, are the solutions of the quadratic equation $x^2 - x - 2 = 0$.

1) $x^2 + 2x - 2 = 0$
2) $x^2 + 2x = 2$
3) $x^2 + 2x + 1 = 2 + 1$
4) $(x + 1)^2 = 3$
5) $x + 1 = \pm\sqrt{3}$
6) $x = -1 \pm \sqrt{3}$

The plus-minus symbol "±" indicates that both $x = -1 + \sqrt{3}$ and $x = -1 - \sqrt{3}$ are solutions of the quadratic equation.

QUADRATIC FORMULA AND ITS DERIVATION

Completing the square can be used to derive a general formula for solving quadratic equations, called the quadratic formula.The mathematical proof will now be briefly summarized. It can easily be seen, by polynomial expansion, that the following equation is equivalent to the quadratic equation:

$$\left(x + \frac{b}{2a}\right)^2 = \frac{b^2 - 4ac}{4a^2}.$$

Taking the square root of both sides, and isolating x, gives:

$$x = \frac{-b \pm \sqrt{b^2 - 4ac}}{2a}.$$

Some sources, particularly older ones, use alternative parameterizations of the quadratic equation such as $ax^2 + 2bx + c = 0$or $ax^2 - 2bx + c = 0$, where b has a magnitude one half of the more common one, possibly with opposite sign. These result in slightly different forms for the solution, but are otherwise equivalent.

A number of alternative derivations can be found in the literature. These proofs are simpler than the standard completing the square method, represent interesting applications of other frequently used techniques in algebra, or offer insight into other areas of mathematics.

A lesser known quadratic formula, as used in Muller's method, and which can be found from Vieta's formulas, provides the same roots via the equation:

$$x = \frac{-2c}{b \pm \sqrt{b^2 - 4ac}}.$$

One property of this form is that it yields one valid root when a = 0, while the other root contains division by zero, because when a = 0, the quadratic equation becomes a linear equation, which has one root. By contrast, in this case the more common formula has a division by zero for one root and an indeterminate form 0/0 for the other root. On the other hand, when c = 0, the more common formula yields two correct roots whereas this form yields the zero root and an indeterminate form 0/0.

In numerical computation, computing with numbers close to zero may lead to large rounding errors. Therefore it is worth to mix the two formulas:

$$x_1 = \frac{-b - \varepsilon\sqrt{b^2 - 4ac}}{2a}, \qquad x_2 = \frac{-2c}{b + \varepsilon\sqrt{b^2 - 4ac}},$$

where $\varepsilon = 1$ if $b \geq 0$, and $\varepsilon = -1$ otherwise.

Reduced Quadratic Equation

It is sometimes convenient to reduce a quadratic equation so that its leading coefficient is one. This is done by dividing both sides by a, which is always possible since a is non-zero. This produces the reduced quadratic equation:

$$x^2 + px + q = 0,$$

where p = b/a and q = c/a. This monic equation has the same solutions as the original.

The quadratic formula for the solutions of the reduced quadratic equation, written in terms of its coefficients, is:

$$x = \frac{1}{2}\left(-p \pm \sqrt{p^2 - 4q}\right).$$

Discriminant

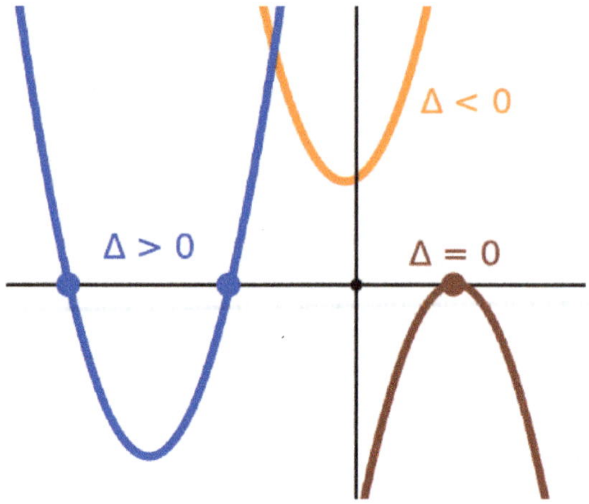

Figure 3. Discriminant signs

In the quadratic formula, the expression underneath the square root sign is called the discriminant of the quadratic equation, and is often represented using an upper case D or an upper case Greek delta:

$$\Delta = b^2 - 4ac.$$

A quadratic equation with real coefficients can have either one or two distinct real roots, or two distinct complex roots. In this case the discriminant determines the number and nature of the roots. There are three cases:

- If the discriminant is positive, then there are two distinct roots

$$\frac{-b+\sqrt{\Delta}}{2a} \quad \text{and} \quad \frac{-b-\sqrt{\Delta}}{2a},$$

both of which are real numbers. For quadratic equations with rational coefficients, if the discriminant is a square number, then the roots are rational—in other cases they may be quadratic irrationals.

- If the discriminant is zero, then there is exactly one real root

$$-\frac{b}{2a},$$

sometimes called a repeated or double root.

- If the discriminant is negative, then there are no real roots. Rather, there are two distinct (non-real) complex roots

$$\frac{-b}{2a}+i\frac{\sqrt{-\Delta}}{2a} \quad \text{and} \quad \frac{-b}{2a}-i\frac{\sqrt{-\Delta}}{2a},$$

which are complex conjugates of each other. In these expressions i is the imaginary unit.

Thus the roots are distinct if and only if the discriminant is non-zero, and the roots are real if and only if the discriminant is non-negative.

Geometric Interpretation

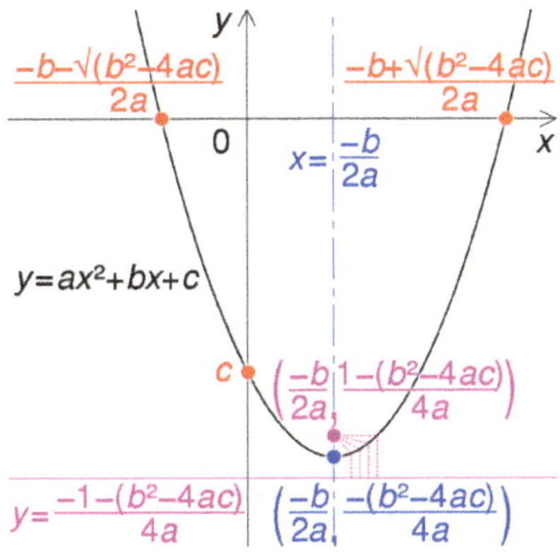

Graph of $y = ax^2 + bx + c$, where a and the discriminant $(b^2 - 4ac)$ are positive, with

- Roots and y-intercept in red
- Vertex and axis of symmetry in blue
- Focus and directrix in pink

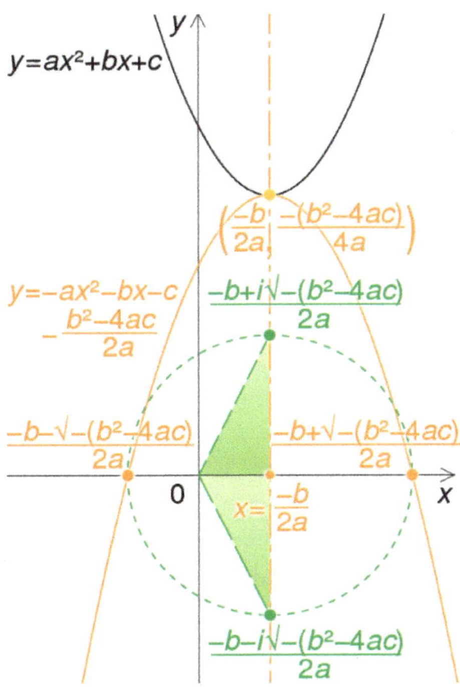

Visualisation of the complex roots of $y = ax^2 + bx + c$: the parabola is rotated 180° about its vertex (yellow). Its x-intercepts are rotated 90° around their mid-point, and the Cartesian plane is interpreted as the complex plane (green).

The function $f(x) = ax^2 + bx + c$ is the quadratic function. The graph of any quadratic function has the same general shape, which is called a parabola. The location and size of the parabola, and how it opens, depend on the values of a, b, and c. As shown in Figure 1, if $a > 0$, the parabola has a minimum point and opens upward. If $a < 0$, the parabola has a maximum point and opens downward. The extreme point of the parabola, whether minimum or maximum, corresponds to its vertex. The x-coordinate of the vertex will be located at $x=\frac{-b}{2a}$ and the y-coordinate of the vertex may be found by substituting this x-value into the function. The y-intercept is located at the point (0, c).

The solutions of the quadratic equation $ax^2 + bx + c = 0$ correspond to the roots of the function $f(x) = ax^2 + bx + c$, since they arc the values of x for which $f(x) = 0$. As shown in Figure 2, if a, b, and c are real numbers and the domain of f is the set of real numbers, then the roots of f are exactly the x-coordinates of the points where the graph touches the x-axis. As shown in Figure 3, if the discriminant is positive, the graph touches the x-axis at two points; if zero, the graph touches at one point; and if negative, the graph does not touch the x-axis.

Quadratic Factorization

The term

$x - r$

is a factor of the polynomial

$$ax^2 + bx + c$$

if and only if r is a root of the quadratic equation

$$ax^2 + bx + c = 0.$$

It follows from the quadratic formula that

$$ax^2 + bx + c = a\left(x - \frac{-b + \sqrt{b^2 - 4ac}}{2a}\right)\left(x - \frac{-b - \sqrt{b^2 - 4ac}}{2a}\right)$$

In the special case $b^2 = 4ac$ where the quadratic has only one distinct root (i.e. the discriminant is zero), the quadratic polynomial can be factored as

$$ax^2 + bx + c = a\left(x + \frac{b}{2a}\right)^2.$$

Graphing for Real Roots

For most of the 20th century, graphing was rarely mentioned as a method for solving quadratic equations in high school or college algebra texts. Students learned to solve quadratic equations by factoring, completing the square, and applying the quadratic formula. Recently, graphing calculators have become common in schools and graphical methods have started to appear in textbooks, but they are generally not highly emphasized.

Being able to use a graphing calculator to solve a quadratic equation requires the ability to produce a graph of y = f(x), the ability to scale the graph appropriately to the dimensions of the graphing surface, and the recognition that when f(x) = 0, x is a solution to the equation. The skills required to solve a quadratic equation on a calculator are in fact applicable to finding the real roots of any arbitrary function.

Since an arbitrary function may cross the x-axis at multiple points, graphing calculators generally require one to identify the desired root by positioning a cursor at a "guessed" value for the root. (Some graphing calculators require bracketing the root on both sides of the zero.) The calculator then proceeds, by an iterative algorithm, to refine the estimated position of the root to the limit of calculator accuracy.

Avoiding Loss of Significance

Although the quadratic formula provides an exact solution, the result is not exact if real numbers are approximated during the computation, as usual in numerical analysis, where real numbers are approximated by floating point numbers (called "reals" in many programming languages). In this context, the quadratic formula is not completely stable.

This occurs when the roots have different order of magnitude, or, equivalently, when b^2 and $b^2 - 4ac$ are close in magnitude. In this case, the subtraction of two nearly equal numbers will cause loss of significance or catastrophic cancellation in the smaller root. To avoid this, the root that is smaller in magnitude, r, can be computed as R where R is the root that is bigger in magnitude.

A second form of cancellation can occur between the terms b^2 and 4ac of the discriminant, that is when the two roots are very close. This can lead to loss of up to half of correct significant figures in the roots.

Examples and Applications

The golden ratio is found as the solution of the quadratic equation

The equations of the circle and the other conic sections—ellipses, parabolas, and hyperbolas—are quadratic equations in two variables.

Given the cosine or sine of an angle, finding the cosine or sine of the angle that is half as large involves solving a quadratic equation.

The process of simplifying expressions involving the square root of an expression involving the square root of another expression involves finding the two solutions of a quadratic equation.

Descartes' theorem states that for every four kissing (mutually tangent) circles, their radii satisfy a particular quadratic equation.

The equation given by Fuss' theorem, giving the relation among the radius of a bicentric quadrilateral's inscribed circle, the radius of its circumscribed circle, and the distance between the centers of those circles, can be expressed as a quadratic equation for which the distance between the two circles' centers in terms of their radii is one of the solutions. The other solution of the same equation in terms of the relevant radii gives the distance between the circumscribed circle's center and the center of the excircle of an ex-tangential quadrilateral.

Babylonian mathematicians, as early as 2000 BC (displayed on Old Babylonian clay tablets) could solve problems relating the areas and sides of rectangles. There is evidence dating this algorithm as far back as the Third Dynasty of Ur. In modern notation, the problems typically involved solving a pair of simultaneous equations of the form:

$$x + y = p, \quad xy = q$$

which are equivalent to the equation:

$$x^2 + q = px$$

The steps given by Babylonian scribes for solving the above rectangle problem were as follows:

Compute half of p.

Square the result.

Subtract q.

Find the square root using a table of squares.

Add together the results of steps (1) and (4) to give x. This is essentially equivalent to calculating

$$x = \frac{p}{2} + \sqrt{\left(\frac{p}{2}\right)^2 - q}$$

Geometric methods were used to solve quadratic equations in Babylonia, Egypt, Greece, China, and India. The Egyptian Berlin Papyrus, dating back to the Middle Kingdom (2050 BC to 1650 BC), contains the solution to a two-term quadratic equation. In the Indian Sulba Sutras, circa 8th century BC, quadratic equations of the form $ax^2 = c$ and $ax^2 + bx = c$ were explored using geometric methods. Babylonian mathematicians from circa 400 BC and Chinese mathematicians from circa 200 BC used geometric methods of dissection to solve quadratic equations with positive roots. Rules for quadratic equations were given in the The Nine Chapters on the Mathematical Art, a Chinese treatise on mathematics. These early geometric methods do not appear to have had a general formula. Euclid, the Greek mathematician, produced a more abstract geometrical method around 300 BC. With a purely geometric approach Pythagoras and Euclid created a general procedure to find solutions of the quadratic equation. In his work Arithmetica, the Greek mathematician Diophantus solved the quadratic equation, but giving only one root, even when both roots were positive.

In 628 AD, Brahmagupta, an Indian mathematician, gave the first explicit (although still not completely general) solution of the quadratic equation $ax^2 + bx = c$ as follows: "To the absolute number multiplied by four times the square, add the square of the middle term; the square root of the same, less the middle term, being divided by twice the square is the value." (Brahmasphutasiddhanta, Colebrook translation, 1817, page 346) This is equivalent to:

$$x = \frac{\sqrt{4ac + b^2} - b}{2a}.$$

The Bakhshali Manuscript written in India in the 7th century AD contained an algebraic formula for solving quadratic equations, as well as quadratic indeterminate equations (originally of type ax/c = y Muhammad ibn Musa al-Khwarizmi (Persia, 9th century), inspired by Brahmagupta, developed a set of formulas that worked for positive solutions. Al-Khwarizmi goes further in providing a full solution to the general quadratic equation, accepting one or two numerical answers for every quadratic equation, while providing geometric proofs in the process. He also described the method of completing the square and recognized that the discriminant must be positive, which was proven by his contemporary 'Abd al-Hamīd ibn Turk (Central Asia, 9th century) who gave geometric figures to prove that if the discriminant is negative, a quadratic equation has no solution. While al-Khwarizmi himself did not accept negative solutions, later Islamic mathematicians that succeeded him accepted negative solutions, as well as irrational numbers as solutions. Abū Kāmil Shujā ibn Aslam (Egypt, 10th century) in particular was the first to accept irrational numbers (often in the form of a square root, cube root or fourth root) as solutions to quadratic equations or as coefficients in an equation. The 9th century Indian mathematician Sridhara wrote down rules for solving quadratic equations.

The Jewish mathematician Abraham bar Hiyya Ha-Nasi (12th century, Spain) authored the first European book to include the full solution to the general quadratic equation. His solution was largely based on Al-Khwarizmi's work. The writing of the Chinese mathematician Yang Hui (1238–1298 AD) is the first known one in which quadratic equations with negative coefficients of ‹x' appear, although he attributes this to the earlier Liu Yi. By 1545 Gerolamo Cardano compiled the works related to the quadratic equations. The quadratic formula covering all cases was first obtained by Simon Stevin in 1594. In 1637 René Descartes published La Géométrie containing the quadratic formula in the form we know today. The first appearance of the general solution in the modern mathematical literature appeared in an 1896 paper by Henry Heaton.

ADVANCED TOPICS

Alternative Methods of Root Calculation

Vieta's Formulas

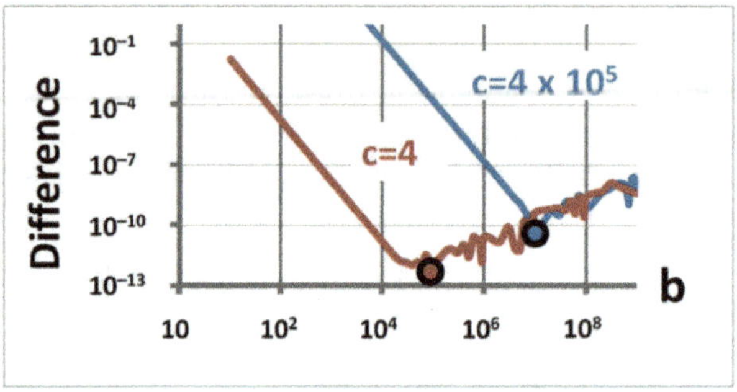

Figure 5. Graph of the difference between Vieta's approximation for the smaller of the two roots of the quadratic equation $x^2 + bx + c = 0$ compared with the value calculated using the quadratic formula. Vieta's approximation is inaccurate for small b but is accurate for large b. The direct evaluation using the quadratic formula is accurate for small bwith roots of comparable value but experiences loss of significance errors for large b and widely spaced roots. The difference between Vieta's approximation versus the direct computation reaches a minimum at the large dots, and rounding causes squiggles in the curves beyond this minimum.

Vieta's formulas give a simple relation between the roots of a polynomial and its coefficients. In the case of the quadratic polynomial, they take the following form:

$$x_1 + x_2 = -\frac{b}{a}$$

and

$$x_1\, x_2 = \frac{c}{a}.$$

These results follow immediately from the relation:

$$(x - x_1)\ (x - x_2) = x^2\ - (x_1 + x_2)\, x + x_1 x_2 = 0,$$

which can be compared term by term with

$$x^2 + (b/a)x + c/a = 0.$$

The first formula above yields a convenient expression when graphing a quadratic function. Since the graph is symmetric with respect to a vertical line through the vertex, when there are two real roots the vertex's x-coordinate is located at the average of the roots (or intercepts). Thus the x-coordinate of the vertex is given by the expression

$$x_V = \frac{x_1 + x_2}{2} = -\frac{b}{2a}.$$

The y-coordinate can be obtained by substituting the above result into the given quadratic equation, giving

$$y_V = -\frac{b^2}{4a} + c = -\frac{b^2 - 4ac}{4a}.$$

As a practical matter, Vieta's formulas provide a useful method for finding the roots of a quadratic in the case where one root is much smaller than the other. If $|x_2| << |x_1|$, then $x_1 + x_2 \approx x_1$, and we have the estimate:

$$x_1 \approx -\frac{b}{a}.$$

The second Vieta's formula then provides:

$$x_2 = \frac{c}{a\,x_1} \approx -\frac{c}{b}.$$

These formulas are much easier to evaluate than the quadratic formula under the condition of one large and one small root, because the quadratic formula evaluates the small root as the difference of two very nearly equal numbers (the case of large b), which causes round-off error in a numerical evaluation. Figure 5 shows the difference between (i) a direct evaluation using the quadratic formula (accurate when the roots are near each other in value) and (ii) an evaluation based upon the above approximation of Vieta's formulas (accurate when the roots are widely spaced). As the linear coefficient b increases, initially the quadratic formula is accurate, and the approximate formula improves in accuracy, leading to a smaller difference between the methods as b increases. However, at some point the quadratic formula begins to lose accuracy because of round off error, while the approximate method continues to improve. Consequently, the difference between the methods begins to increase as the quadratic formula becomes worse and worse.

This situation arises commonly in amplifier design, where widely separated roots are desired to ensure a stable operation (see step response).

Trigonometric Solution

In the days before calculators, people would use mathematical tables—lists of numbers showing the results of calculation with varying arguments—to simplify and speed up computation. Tables of logarithms and trigonometric functions were common in math and science textbooks. Specialized tables were published for applications such as astronomy, celestial navigation and statistics. Methods of numerical approximation existed, called prosthaphaeresis, that offered shortcuts around time-consuming operations such as multiplication and taking powers and roots. Astronomers, especially, were concerned with methods that could speed up the long series of computations involved in celestial mechanics calculations.

It is within this context that we may understand the development of means of solving quadratic equations by the aid of trigonometric substitution. Consider the following alternate form of the quadratic equation,

[1] $ax^2 + bx \pm c = 0,$

where the sign of the ± symbol is chosen so that a and c may both be positive.

By substituting

[2] $x = \sqrt{c/a}\tan\theta$

and then multiplying through by $\cos^2\theta$, we obtain

[3] $\sin^2\theta + \frac{b}{\sqrt{ac}}\sin\theta\cos\theta \pm \cos^2\theta = 0.$

Introducing functions of 2θ and rearranging, we obtain

[4] $\tan 2\theta_n = +2\frac{\sqrt{ac}}{b},$

[5] $\sin 2\theta_p = -2\frac{\sqrt{ac}}{b},$

where the subscripts n and p correspond, respectively, to the use of a negative or positive sign in equation . Substituting the two values of θ_n or θ_p found from equations or into gives the required roots of . Complex roots occur in the solution based on equation if the absolute value of sin $2\theta_p$ exceeds unity. The amount of effort involved in solving quadratic equations using this mixed trigonometric and logarithmic table look-up strategy was two-thirds the effort using logarithmic tables alone. Calculating complex roots would require using a different trigonometric form.

To illustrate, let us assume we had available seven-place logarithm and trigonometric tables, and wished to solve the following to six-significant-figure accuracy:

$$4.16130x^2 + 9.15933x - 11.4207 = 0$$

1. A seven-place lookup table might have only 100,000 entries, and computing intermediate results to seven places would generally require interpolation between adjacent entries.
2. $\log a = 0.6192290, \log b = 0.9618637, \log c = 1.0576927$
3. $2\sqrt{ac}/b = 2 \times 10^{(0.6192290+1.0576927)/2-0.9618637} = 1.505314$
4. $\theta = (\tan^{-1} 1.505314)/2 = 28.20169^\circ$ or -61.79831°
5. $\log|\tan\theta| = -0.2706462$ or 0.2706462
6. $\log\sqrt{c/a} = (1.0576927 - 0.6192290)/2 = 0.2192318$
7. $x_1 = 10^{0.2192318-0.2706462} = 0.888353$ (rounded to six significant figures)

   $x_2 = -10^{0.2192318+0.2706462} = -3.08943$

Solution for Complex Roots in Polar Coordinates

If the quadratic equation $ax^2 + bx + c = 0$ with real coefficients has two complex roots—the case where $b^2 - 4ac < 0$, , requiring a and c to have the same sign as each other—then the solutions for the roots can be expressed in polar form as

$$x_1,\ x_2 = r(\cos\theta \pm i\sin\theta),$$

where $r = \sqrt{\frac{c}{a}}$ and $\theta = \cos^{-1}\left(\frac{-b}{2\sqrt{ac}}\right)$.

Geometric Solution

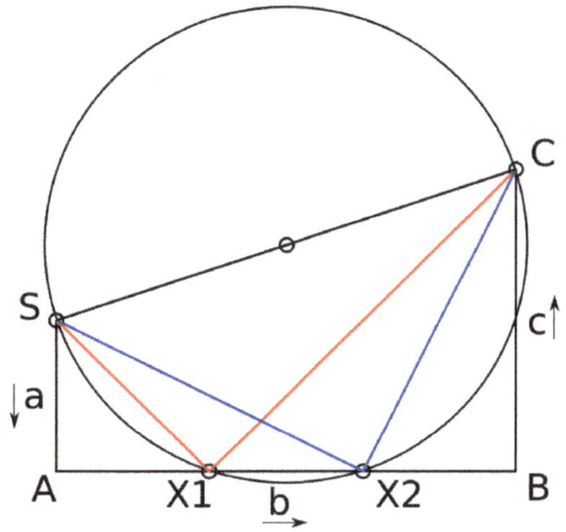

Figure 6. Geometric solution of $ax^2 + bx + c = 0$ using Lill's method. Solutions are −AX1/SA, −AX2/SA

The quadratic equation may be solved geometrically in a number of ways. One way is via Lill's method. The three coefficients a, b, c are drawn with right angles between them as in SA, AB, and BC in Figure 6. A circle is drawn with the start and end point SC as a diameter. If this cuts the middle line AB of the three then the equation has a solution, and the solutions are given by negative of the distance along this line from A divided by the first coefficient a or SA. If a is 1 the coefficients may be read off directly. Thus the solutions in the diagram are −AX1/SA and −AX2/SA.

The Carlyle circle, named after Thomas Carlyle, has the property that the solutions of the quadratic equation are the horizontal coordinates of the intersections of the circle with the horizontal axis. Carlyle circles have been used to develop ruler-and-compass constructions of regular polygons.

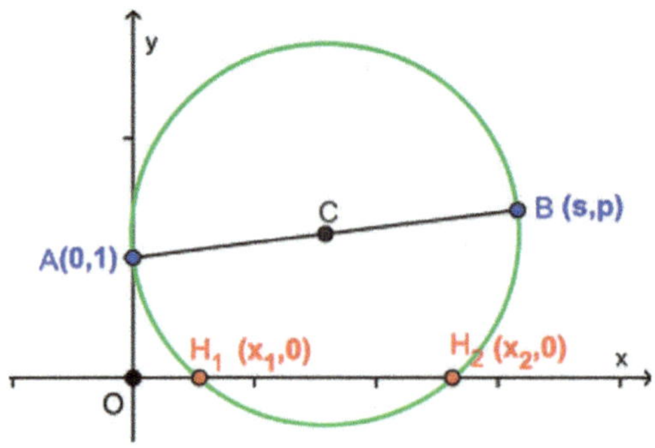

Carlyle circle of the quadratic equation $x^2 - sx + p = 0$.

Generalization of Quadratic Equation

The formula and its derivation remain correct if the coefficients a, b and care complex numbers, or more generally members of any field whose characteristic is not 2. (In a field of characteristic 2, the element 2a is zero and it is impossible to divide by it.)

The symbol

$$\pm\sqrt{b^2 - 4ac}$$

in the formula should be understood as "either of the two elements whose square is $b^2 - 4ac$, if such elements exist". In some fields, some elements have no square roots and some have two; only zero has just one square root, except in fields of characteristic 2. Even if a field does not contain a square root of some number, there is always a quadratic extension field which does, so the quadratic formula will always make sense as a formula in that extension field.

Characteristic 2

In a field of characteristic 2, the quadratic formula, which relies on 2 being a unit, does not hold. Consider the monic quadratic polynomial

$$x^2 + bx + c$$

over a field of characteristic 2. If $b = 0$, then the solution reduces to extracting a square root, so the solution is

$$x = \sqrt{c}$$

and there is only one root since

$$-\sqrt{c} = -\sqrt{c} + 2\sqrt{c} = \sqrt{c}.$$

In summary,

$$x^2 + c = (x + \sqrt{c})^2.$$

See quadratic residue for more information about extracting square roots in finite fields.

In the case that $b \neq 0$, there are two distinct roots, but if the polynomial is irreducible, they cannot be expressed in terms of square roots of numbers in the coefficient field. Instead, define the **2-root** R(c) of c to be a root of the polynomial $x^2 + x + c$, an element of the splitting field of that polynomial. One verifies that R(c) + 1 is also a root. In terms of the 2-root operation, the two roots of the (non-monic) quadratic $ax^2 + bx + c$ are

$$\frac{b}{a}R\left(\frac{ac}{b^2}\right)$$

and

$$\frac{b}{a}\left(R\left(\frac{ac}{b^2}\right)+1\right).$$

For example, let a denote a multiplicative generator of the group of units of F_4, the Galois field of order four (thus a and a + 1are roots of $x^2 + x + 1$ over F_4. Because $(a + 1)^2 = a$, a + 1 is the unique solution of the quadratic equation $x^2 + a = 0$. On the other hand, the polynomial $x^2 + ax + 1$ is irreducible over F_4, but it splits over F_{16}, where it has the two roots ab and ab + a, where b is a root of $x^2 + x + a$ in F_{16}.

This is a special case of Artin–Schreier theory.

REFERENCES

1. Protters & Morrey: " Calculus and Analytic Geometry. First Course"
2. Washington, Allyn J. (2000). Basic Technical Mathematics with Calculus, Seventh Edition. Addison Wesley Longman, Inc. ISBN 0-201-35666-X.
3. Ebbinghaus, Heinz-Dieter; Ewing, John H. (1991), Numbers, Graduate Texts in Mathematics, 123, Springer, p. 77, ISBN 9780387974972.
4. Sterling, Mary Jane (2010), Algebra I For Dummies, Wiley Publishing, p. 219, ISBN 978-0-470-55964-2
5. Rich, Barnett; Schmidt, Philip (2004), Schaum's Outline of Theory and Problems of Elementary Algebra, The McGraw-Hill Companies, ISBN 0-07-141083-X, Chapter 13 §4.4, p. 291
6. Himonas, Alex. Calculus for Business and Social Sciences, p. 64 (Richard Dennis Publications, 2001).
7. Kahan, Willian (November 20, 2004), On the Cost of Floating-Point Computation Without Extra-Precise Arithmetic (PDF), retrieved 2012-12-25
8. Alenit□syn, Aleksandr and Butikov, Evgeniĭ. Concise Handbook of Mathematics and Physics, p. 38 (CRC Press 1997)
9. Δ is the initial of the Greek word Διακρίνουσα, Diakrínousa, discriminant.
10. Achatz, Thomas; Anderson, John G.; McKenzie, Kathleen (2005). Technical Shop Mathematics. Industrial Press. p. 277. ISBN 0-8311-3086-5.
11. "Complex Roots Made Visible -- Math Fun Facts". Retrieved 1 October 2016.
12. Wharton, P. (2006). Essentials of Edexcel Gcse Math/Higher. Lonsdale. p. 63. ISBN 978-1-905-129-78-2.
13. Ballew, Pat. "Solving Quadratic Equations — By analytic and graphic methods;

Including several methods you may never have seen" (PDF). Retrieved 18 April 2013.

14. Higham, Nicholas (2002), Accuracy and Stability of Numerical Algorithms (2nd ed.), SIAM, p. 10, ISBN 978-0-89871-521-7
15. Friberg, Jöran (2009). "A Geometric Algorithm with Solutions to Quadratic Equations in a Sumerian Juridical Document from Ur III Umma". Cuneiform Digital Library Journal. 3.
16. Stillwell, John (2004). Mathematics and Its History (2nd ed.). Springer. ISBN 0-387-95336-1.
17. The Cambridge Ancient History Part 2 Early History of the Middle East. Cambridge University Press. 1971. p. 530. ISBN 978-0-521-07791-0.
18. Henderson, David W. "Geometric Solutions of Quadratic and Cubic Equations". Mathematics Department, Cornell University. Retrieved 28 April 2013.
19. Aitken, Wayne. "A Chinese Classic: The Nine Chapters" (PDF). Mathematics Department, California State University. Retrieved 28 April 2013.
20. Smith, David Eugene (1958). History of Mathematics. Courier Dover Publications. p. 380. ISBN 978-0-486-20430-7.
21. Smith, David Eugene (1958). History of Mathematics, Volume 1. Courier Dover Publications. p. 134. ISBN 0-486-20429-4. Extract of page 134
22. Katz, V. J.; Barton, B. (2006). "Stages in the History of Algebra with Implications for Teaching". Educational Studies in Mathematics. 66 (2): 185–201. doi:10.1007/s10649-006-9023-7.
23. Boyer, Carl B.; Uta C. Merzbach, rev. editor (1991). A History of Mathematics. John Wiley & Sons, Inc. ISBN 0-471-54397-7.
24. O'Connor, John J.; Robertson, Edmund F., "Arabic mathematics: forgotten brilliance?", MacTutor History of Mathematics archive, University of St Andrews. "Algebra was a unifying theory which allowed rational numbers, irrational numbers, geometrical magnitudes, etc., to all be treated as "algebraic objects"."
25. Jacques Sesiano, "Islamic mathematics", p. 148, in Selin, Helaine; D'Ambrosio, Ubiratan, eds. (2000), Mathematics Across Cultures: The History of Non-Western Mathematics, Springer, ISBN 1-4020-0260-2
26. Smith, David Eugene (1958). History of Mathematics. Courier Dover Publications. p. 280. ISBN 978-0-486-20429-1.
27. Livio, Mario (2006). The Equation that Couldn't Be Solved. Simon & Schuster. ISBN 0743258215.
28. Ronan, Colin (1985). The Shorter Science and Civilisation in China. Cambridge University Press. p. 15. ISBN 978-0-521-31536-4.
29. Struik, D. J.; Stevin, Simon (1958), The Principal Works of Simon Stevin, Mathematics (PDF), II–B, C. V. Swets & Zeitlinger, p. 470
30. Heaton, H (1896). "A Method of Solving Quadratic Equations". American Mathematical Monthly. 3 (10): 236–237. doi:10.2307/2971099. JSTOR 2971099.

31. Seares, F. H. (1945). "Trigonometric Solution of the Quadratic Equation". Publications of the Astronomical Society of the Pacific. 57 (339): 307–309. Bibcode:1945PASP...57..307S. doi:10.1086/125759.
32. Aude, H. T. R. (1938). "The Solutions of the Quadratic Equation Obtained by the Aid of the Trigonometry". National Mathematics Magazine. 13 (3): 118–121. doi:10.2307/3028750. JSTOR 3028750.
33. Simons, Stuart, "Alternative approach to complex roots of real quadratic equations", Mathematical Gazette 93, March 2009, 91–92.
34. Bixby, William Herbert (1879), Graphical Method for finding readily the Real Roots of Numerical Equations of Any Degree, West Point N. Y.
35. Weisstein, Eric W. "Carlyle Circle". From MathWorld—A Wolfram Web Resource. Retrieved 21 May 2013.

Linear Algebra

Chapter Contents

Linear algebra is the study of linear sets of equations and their transformation properties. Linear algebra allows the analysis of rotationsin space, least squares fitting, solution of coupled differential equations, determination of a circle passing through three given points, as well as many other problems in mathematics, physics, and engineering. Confusingly, linear algebra is not actually an algebra in the technical sense of the word "algebra" (i.e., a vector space V over a field F, and so on).

INTRODUCTION

Linear algebra is the branch of mathematics concerning vector spaces and linear mappings between such spaces. It includes the study of lines, planes, and subspaces, but is also concerned with properties common to all vector spaces.

The set of points with coordinates that satisfy a linear equation forms a hyperplane in an n-dimensional space. The conditions under which a set of n hyperplanes intersect in a single point is an important focus of study in linear algebra. Such an investigation is initially motivated by a system of linear equations containing several unknowns. Such equations are naturally represented using the formalism of matrices and vectors.

Linear algebra is central to both pure and applied mathematics. For instance, abstract algebra arises by relaxing the axioms of a vector space, leading to a number of generalizations. Functional analysis studies the infinite-dimensional version of the theory of vector spaces. Combined with calculus, linear algebra facilitates the solution of linear systems of differential equations.

Techniques from linear algebra are also used in analytic geometry, engineering, physics, natural sciences, computer science, computer animation, advanced facial recognition algorithms and the social sciences (particularly in economics). Because linear algebra is such a well-developed theory, nonlinear mathematical models are sometimes approximated by linear models.

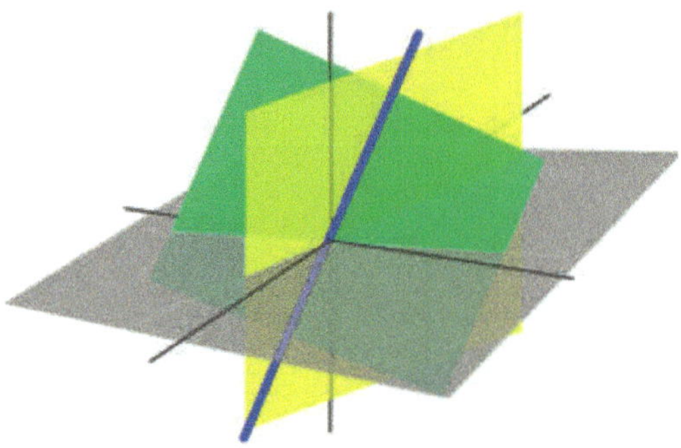

The three-dimensional Euclidean space$\mathbf{R}^3$ is a vector space, and lines and planes passing through the origin are vector subspaces in $\mathbf{R}^3$.

The study of linear algebra first emerged from the study of determinants, which were used to solve systems of linear equations. Determinants were used by Leibniz in 1693, and subsequently, Gabriel Cramer devised Cramer's Rule for solving linear systems in 1750. Later, Gauss further developed the theory of solving linear systems by using Gaussian elimination, which was initially listed as an advancement in geodesy.

The study of matrix algebra first emerged in England in the mid-1800s. In 1844 Hermann Grassmann published his "Theory of Extension" which included foundational new topics of what is today called linear algebra. In 1848, James Joseph Sylvester introduced the term matrix, which is Latin for "womb". While studying compositions of linear transformations, Arthur Cayley was led to define matrix multiplication and inverses. Crucially, Cayley used a single letter to denote a matrix, thus treating a matrix as an aggregate object. He also realized the connection between matrices and determinants, and wrote "There would be many things to say about this theory of matrices which should, it seems to me, precede the theory of determinants".

In 1882, Hüseyin Tevfik Pasha wrote the book titled "Linear Algebra". The first modern and more precise definition of a vector space was introduced by Peano in 1888; by 1900, a theory of linear transformations of finite-dimensional vector spaces had emerged. Linear algebra took its modern form in the first half of the twentieth century, when many ideas and methods of previous centuries were generalized as abstract algebra. The use of matrices in quantum mechanics, special relativity, and statistics helped spread the subject of linear algebra beyond pure mathematics. The development of computers led to increased research in efficient algorithms for Gaussian elimination and matrix decompositions, and linear algebra became an essential tool for modelling and simulations.

The origin of many of these ideas is discussed in the articles on determinants and Gaussian elimination.

EDUCATIONAL HISTORY

Linear algebra first appeared in American graduate textbooks in the 1940s and in undergraduate textbooks in the 1950s. Following work by the School Mathematics Study Group, U.S. high schools asked 12th grade students to do "matrix algebra, formerly reserved for college" in the 1960s. In France during the 1960s, educators attempted to teach linear algebra through affine dimensional vector spaces in the first year of secondary school. This was met with a backlash in the 1980s that removed linear algebra from the curriculum. In 1993, the U.S.-based Linear Algebra Curriculum Study Group recommended that undergraduate linear algebra courses be given an application-based "matrix orientation" as opposed to a theoretical orientation. Reviews of the teaching of linear algebra call for stress on visualization and geometric interpretation of theoretical ideas, and to include the jewel in the crown of linear algebra, the singular value decomposition (SVD), as ‹so many other disciplines use it'. To better suit 21st century applications, such as data mining and uncertainty analysis, linear algebra can be based upon the SVD instead of Gaussian Elimination

SCOPE OF STUDY

Vector Spaces

The main structures of linear algebra are vector spaces. A vector space over a field F is a set V together with two binary operations. Elements of V are called vectors and elements of F are called scalars. The first operation, vector addition, takes any two vectors v and w and outputs a third vector v + w. The second operation, scalar multiplication, takes any scalar a and any vector v and outputs a new vector av. The operations of addition and multiplication in a vector space must satisfy the following axioms. In the list below, let u, v and w be arbitrary vectors in V, and a and b scalars in F.

Axiom	Meaning
Associativity of addition	u + (v + w) = (u + v) + w
Commutativity of addition	u + v = v + u
Identity element of addition	There exists an element 0 ε V, called the zero vector, such that v + 0 = v for all v ε V.

Inverse elements of addition	For every v ε V, there exists an element −v ε V, called the additive inverse of v, such that v + (−v) = 0.
Compatibility of scalar multiplication with field multiplication	a(bv) = (ab)v
Identity element of scalar multiplication	1v = v, where 1 denotes the multiplicative identity in F.
Distributivity of scalar multiplication with respect to vector addition	a(u + v) = au + av
Distributivity of scalar multiplication with respect to field addition	(a + b)v = av + bv

The first four axioms are those of V being an abelian group under vector addition. Vector spaces may be diverse in nature, for example, containing functions, polynomials or matrices. Linear algebra is concerned with properties common to all vector spaces.

Linear Transformations

Similarly as in the theory of other algebraic structures, linear algebra studies mappings between vector spaces that preserve the vector-space structure. Given two vector spaces V and W over a field **F**, a linear transformation (also called linear map, linear mapping or linear operator) is a map

$$T : V \to W$$

that is compatible with addition and scalar multiplication:

$$T(u+v) = T(u) + T(v), \quad T(av) = aT(v)$$

for any vectors $u,v \in V$ and a scalar $a \in \mathbf{F}$.

Additionally for any vectors $u, v \in V$ and scalars $a, b \in \mathbf{F}$:

$$T(au+bv) = T(au) + T(bv) = aT(u) + bT(v)$$

When a bijective linear mapping exists between two vector spaces (that is, every vector from the second space is associated with exactly one in the first), we say that the two spaces are isomorphic. Because an isomorphism preserves linear structure, two isomorphic vector spaces are "essentially the same" from the linear algebra point of view. One essential question in linear algebra is whether a mapping is an isomorphism or not, and this question can be answered by checking if the determinant is nonzero. If a mapping is not an isomorphism, linear algebra is interested in finding its range (or image) and the set of elements that get mapped to zero, called the kernel of the mapping.

Linear transformations have geometric significance. For example, 2 × 2 real matrices denote standard planar mappings that preserve the origin.

Subspaces, Span, and Basis

Again, in analogue with theories of other algebraic objects, linear algebra is interested in subsets of vector spaces that are themselves vector spaces; these subsets are called linear subspaces. For example, both the range and kernel of a linear mapping are subspaces, and are thus often called the range space and the nullspace; these are important examples of subspaces. Another important way of forming a subspace is to take a linear combination of a set of vectors $v_1, v_2, \ldots, v_k$:

$$a_1 v_1 + a_2 v_2 + \cdots + a_k v_k,$$

where $a_1, a_2, \ldots, a_k$ are scalars. The set of all linear combinations of vectors $v_1, v_2, \ldots, v_k$ is called their span, which forms a subspace.

A linear combination of any system of vectors with all zero coefficients is the zero vector of V. If this is the only way to express the zero vector as a linear combination of $v_1, v_2, \ldots, v_k$ then these vectors are linearly independent. Given a set of vectors that span a space, if any vector w is a linear combination of other vectors (and so the set is not linearly independent), then the span would remain the same if we remove w from the set. Thus, a set of linearly dependent vectors is redundant in the sense that there will be a linearly independent subset which will span the same subspace. Therefore, we are mostly interested in a linearly independent set of vectors that spans a vector space V, which we call a basis of V. Any set of vectors that spans V contains a basis, and any linearly independent set of vectors in V can be extended to a basis. It turns out that if we accept the axiom of choice, every vector space has a basis; nevertheless, this basis may be unnatural, and indeed, may not even be constructible. For instance, there exists a basis for the real numbers, considered as a vector space over the rationals, but no explicit basis has been constructed.

Any two bases of a vector space V have the same cardinality, which is called the dimension of V. The dimension of a vector space is well-defined by the dimension theorem for vector spaces. If a basis of V has finite number of elements, V is called a finite-dimensional vector space. If V is finite-dimensional and U is a subspace of V, then dim U $\leq$ dim V. If U_1 and U_2 are subspaces of V, then

$$\dim(U_1 + U_2) = \dim U_1 + \dim U_2 - \dim(U_1 \cap U_2)$$

One often restricts consideration to finite-dimensional vector spaces. A fundamental theorem of linear algebra states that all vector spaces of the same dimension are isomorphic, giving an easy way of characterizing isomorphism.

MATRIX THEORY

A particular basis $\{v_1, v_2, \ldots, v_n\}$ of V allows one to construct a coordinate system in V: the vector with coordinates $(a_1, a_2, \ldots, a_n)$ is the linear combination

$$a_1 v_1 + a_2 v_2 + \cdots + a_n v_n.$$

The condition that $v_1, v_2, ..., v_n$ span V guarantees that each vector v can be assigned coordinates, whereas the linear independence of $v_1, v_2, ..., v_n$ assures that these coordinates are unique (i.e. there is only one linear combination of the basis vectors that is equal to v). In this way, once a basis of a vector space V over **F** has been chosen, V may be identified with the coordinate n-space $\mathbf{F}^n$. Under this identification, addition and scalar multiplication of vectors in V correspond to addition and scalar multiplication of their coordinate vectors in $\mathbf{F}^n$. Furthermore, if V and W are an n-dimensional and m-dimensional vector space over **F**, and a basis of V and a basis of W have been fixed, then any linear transformation T: V → W may be encoded by an m × n matrix A with entries in the field **F**, called the matrix of T with respect to these bases. Two matrices that encode the same linear transformation in different bases are called similar. Matrix theory replaces the study of linear transformations, which were defined axiomatically, by the study of matrices, which are concrete objects. This major technique distinguishes linear algebra from theories of other algebraic structures, which usually cannot be parameterized so concretely.

There is an important distinction between the coordinate n-space $\mathbf{R}^n$ and a general finite-dimensional vector space V. While $\mathbf{R}^n$ has a standard basis $\{e_1, e_2, ..., e_n\}$, a vector space V typically does not come equipped with such a basis and many different bases exist (although they all consist of the same number of elements equal to the dimension of V).

One major application of the matrix theory is calculation of determinants, a central concept in linear algebra. While determinants could be defined in a basis-free manner, they are usually introduced via a specific representation of the mapping; the value of the determinant does not depend on the specific basis. It turns out that a mapping has an inverse if and only if the determinant has an inverse (every non-zero real or complex number has an inverse). If the determinant is zero, then the nullspace is nontrivial. Determinants have other applications, including a systematic way of seeing if a set of vectors is linearly independent (we write the vectors as the columns of a matrix, and if the determinant of that matrix is zero, the vectors are linearly dependent). Determinants could also be used to solve systems of linear equations (see Cramer's rule), but in real applications, Gaussian elimination is a faster method.

Eigenvalues and Eigenvectors

In general, the action of a linear transformation may be quite complex. Attention to low-dimensional examples gives an indication of the variety of their types. One strategy for a general n-dimensional transformation T is to find "characteristic lines" that are invariant sets under T. If v is a non-zero vector such that Tv is a scalar multiple of v, then the line through 0 and v is an invariant set under T and v is called a characteristic vector or **eigenvector**. The scalar λ such that Tv = λv is called a characteristic value or **eigenvalue** of T.

To find an eigenvector or an eigenvalue, we note that

$$Tv - \lambda v = (T - \lambda \mathrm{I})v = 0,$$

where I is the identity matrix. For there to be nontrivial solutions to that equation, det(T − λ I) = 0. The determinant is a polynomial, and so the eigenvalues are not guaranteed to exist if the field is **R**. Thus, we often work with an algebraically closed field such as the complex numbers when dealing with eigenvectors and eigenvalues so that an eigenvalue will always exist. It would be particularly nice if given a transformation T taking a vector space V into itself we can find a basis for V consisting of eigenvectors. If such a basis exists, we can easily compute the action of the transformation on any vector: if v_1, v_2, ..., v_n are linearly independent eigenvectors of a mapping of n-dimensional spaces T with (not necessarily distinct) eigenvalues λ_1, λ_2, ..., λ_n, and if $v = a_1v_1 + ... + a_n v_n$, then,

$$T(v) = T(a_1v_1) + \cdots + T(a_nv_n) = a_1T(v_1) + \cdots + a_nT(v_n) = a_1\lambda_1v_1 + \cdots + a_n\lambda_nv_n.$$

Such a transformation is called a diagonalizable matrix since in the eigenbasis, the transformation is represented by a diagonal matrix. Because operations like matrix multiplication, matrix inversion, and determinant calculation are simple on diagonal matrices, computations involving matrices are much simpler if we can bring the matrix to a diagonal form. Not all matrices are diagonalizable (even over an algebraically closed field).

Inner-Product Spaces

Besides these basic concepts, linear algebra also studies vector spaces with additional structure, such as an inner product. The inner product is an example of a bilinear form, and it gives the vector space a geometric structure by allowing for the definition of length and angles. Formally, an inner product is a map

- Conjugate symmetry:

$$\langle u, v\rangle = \overline{\langle v, u\rangle}.$$

Note that in **R**, it is symmetric.

- Linearity in the first argument:

$$\langle au, v\rangle = a\langle u, v\rangle.$$
$$\langle u + v, w\rangle = \langle u, w\rangle + \langle v, w\rangle.$$

- Positive-definiteness:

$\langle v, v\rangle \geq 0$ with equality only for $v = 0$.

We can define the length of a vector v in V by

$$\|v\|^2 = \langle v, v\rangle,$$

and we can prove the Cauchy–Schwarz inequality:

$$|\langle u, v\rangle| \leq \|u\| \cdot \|v\|.$$

In particular, the quantity

$$\frac{|\langle u, v\rangle|}{\|u\| \cdot \|v\|} \leq 1,$$

and so we can call this quantity the cosine of the angle between the two vectors.

Two vectors are orthogonal if $\langle u, v \rangle = 0$ An orthonormal basis is a basis where all basis vectors have length 1 and are orthogonal to each other. Given any finite-dimensional vector space, an orthonormal basis could be found by the Gram–Schmidt procedure. Orthonormal bases are particularly nice to deal with, since if v = $a_1 v_1 + \ldots + a_n v_n$, then

$$a_i = \langle v, v_i \rangle$$

The inner product facilitates the construction of many useful concepts. For instance, given a transform T, we can define its Hermitian conjugate T* as the linear transform satisfying

$$\langle Tu, v \rangle = \langle u, T^* v \rangle.$$

If T satisfies TT* = T*T, we call T normal. It turns out that normal matrices are precisely the matrices that have an orthonormal system of eigenvectors that span V.

Some Main Useful Theorems

- A matrix is invertible, or non-singular, if and only if the linear map represented by the matrix is an isomorphism.
- Any vector space over a field **F** of dimension n is isomorphic to $\mathbf{F}^n$ as a vector space over **F**.
- Corollary: Any two vector spaces over **F** of the same finite dimension are isomorphic to each other.
- A linear map is an isomorphism if and only if the determinant is nonzero.

APPLICATIONS

Because of the ubiquity of vector spaces, linear algebra is used in many fields of mathematics, natural sciences, computer science, and social science. Below are just some examples of applications of linear algebra.

Solution of Linear Systems

Linear algebra provides the formal setting for the linear combination of equations used in the Gaussian method. Suppose the goal is to find and describe the solution(s), if any, of the following system of linear equations:

$$\begin{aligned} 2x + y - z &= 8 \quad (L_1) \\ -3x - y + 2z &= -11 \quad (L_2) \\ -2x + y + 2z &= -3 \quad (L_3) \end{aligned}$$

The Gaussian-elimination algorithm is as follows: eliminate x from all equations below L_1, and then eliminate y from all equations below L_2. This will put the system into triangular form. Then, using back-substitution, each unknown can be solved for.

In the example, x is eliminated from L_2 by adding $(3/2)L_1$ to L_2. x is then eliminated from L_3 by adding L_1 to L_3. Formally:

$$L_2 + \tfrac{3}{2}L_1 \to L_2$$
$$L_3 + L_1 \to L_3$$

The result is:

$$\begin{aligned} 2x + y - z &= 8 \\ \frac{1}{2}y + \frac{1}{2}z &= 1 \\ 2y + z &= 5 \end{aligned}$$

Now y is eliminated from L_3 by adding $-4L_2$ to L_3:

$$L_3 + -4L_2 \to L_3$$

The result is:

$$\begin{aligned} 2x + y - z &= 8 \\ \frac{1}{2}y + \frac{1}{2}z &= 1 \\ -z &= 1 \end{aligned}$$

This result is a system of linear equations in triangular form, and so the first part of the algorithm is complete.

The last part, back-substitution, consists of solving for the known in reverse order. It can thus be seen that

z = – 1 (L_3)

Then, z can be substituted into L2, which can then be solved to obtain

y = 3 (L_2)

Next, z and y can be substituted into L1, which can be solved to obtain

x = 2 (L_1)

The system is solved.

We can, in general, write any system of linear equations as a matrix equation:

$A_x = b$.

The solution of this system is characterized as follows: first, we find a particular solution x_0 of this equation using Gaussian elimination. Then, we compute the solutions of Ax = 0; that is, we find the null space N of A. The solution set of this equation is given by $x_0 + N = \{x_0 + n : n \in N\}$. If the number of variables is equal to the number of equations, then we can characterize when the system has a unique solution: since N is trivial if and only if det A ≠ 0, the equation has a unique solution if and only if det A ≠ 0.

Least-Squares Best Fit Line

The least squares method is used to determine the best fit line for a set of data. This line will minimize the sum of the squares of the residuals.

Fourier Series Expansion

Fourier series are a representation of a function f: [−π, π] → **R** as a trigonometric series:

$$f(x) = \frac{a_0}{2} + \sum_{n=1}^{\infty} [a_n \cos(nx) + b_n \sin(nx)].$$

This series expansion is extremely useful in solving partial differential equations. In this article, we will not be concerned with convergence issues; it is nice to note that all Lipschitz-continuous functions have a converging Fourier series expansion, and nice enough discontinuous functions have a Fourier series that converges to the function value at most points.

The space of all functions that can be represented by a Fourier series form a vector space (technically speaking, we call functions that have the same Fourier series expansion the "same" function, since two different discontinuous functions might have the same Fourier series). Moreover, this space is also an inner product space with the inner product

$$\langle f, g \rangle = \frac{1}{\pi} \int_{-\pi}^{\pi} f(x) g(x)\, dx.$$

The functions gn(x) = sin(nx) for n > 0 and hn(x) = cos(nx) for n ≥ 0 are an orthonormal basis for the space of Fourier-expandable functions. We can thus use the tools of linear algebra to find the expansion of any function in this space in terms of these basis functions. For instance, to find the coefficient ak, we take the inner product with hk:

$$\langle f, h_k \rangle = \frac{a_0}{2} \langle h_0, h_k \rangle + \sum_{n=1}^{\infty} [a_n \langle h_n, h_k \rangle + b_n \langle\, g_n, h_k \rangle],$$

and by orthonormality, $\langle f, h_k \rangle = a_k$; that is,

$$a_k = \frac{1}{\pi} \int_{-\pi}^{\pi} f(x) \cos(kx)\, dx.$$

Quantum Mechanics

Quantum mechanics is highly inspired by notions in linear algebra. In quantum mechanics, the physical state of a particle is represented by a vector, and observables (such as momentum, energy, and angular momentum) are represented by linear operators on the underlying vector space. More concretely, the wave function of a particle describes its physical state and lies in the vector space L^2 (the functions φ: $\mathbf{R}^3$ → **C** such $\int_{-\infty}^{\infty}\int_{-\infty}^{\infty}\int_{-\infty}^{\infty} |\phi|^2 dxdydz$ finite), and it evolves according to the Schrödinger equation. Energy is represented as the operator $H = -\frac{\hbar^2}{2m}\nabla^2 + V(x, y, z)$, where V is the potential energy. H is also known as the Hamiltonian operator. The eigenvalues of H represents the possible energies that can be observed. Given a particle in some state φ, we can expand φ into a linear combination of eigenstates of H. The component of H in each eigenstate determines the probability of measuring the corresponding eigenvalue, and the measurement forces the particle to assume that eigenstate (wave function collapse).

Geometric Introduction

Many of the principles and techniques of linear algebra can be seen in the geometry of lines in a real two dimensional plane E. When formulated using vectors and matrices the geometry of points and lines in the plane can be extended to the geometry of points and hyperplanes in high-dimensional spaces.

Point coordinates in the plane E are ordered pairs of real numbers, (x,y), and a line is defined as the set of points (x,y) that satisfy the linear equation

$$\lambda : ax + by + c = 0,$$

where *a*, *b* and *c* are not all zero. Then,

$$\lambda : [a \quad b \quad c]\begin{Bmatrix} x \\ y \\ 1 \end{Bmatrix} = 0,$$

or

$$A\mathbf{x} = 0,$$

where $\mathbf{x} = (x, y, 1)$ is the 3×1 set of homogeneous coordinates associated with the point (x, y).

Homogeneous coordinates identify the plane E with the z = 1 plane in three dimensional space. The x–y coordinates in E are obtained from homogeneous coordinates $\mathbf{y} = (y_1, y_2, y_3)$ by dividing by the third component (if it is nonzero) to obtain $\mathbf{y} = (y_1/y_3, y_2/y_3, 1)$.

The linear equation, λ, has the important property, that if $\mathbf{x}_1$ and $\mathbf{x}_2$ are homogeneous coordinates of points on the line, then the point $\alpha\mathbf{x}_1 + \beta\mathbf{x}_2$ is also on the line, for any real α and β.

Now consider the equations of the two lines λ_1 and λ_2,

$$\lambda_1 : a_1x + b_1y + c_1 = 0, \quad \lambda_2 : a_2x + b_2y + c_2 = 0,$$

which forms a system of linear equations. The intersection of these two lines is defined by x = (x, y, 1) that satisfy the matrix equation,

$$\lambda_{1,2} : \begin{bmatrix} a_1 & b_1 & c_1 \\ a_2 & b_2 & c_2 \end{bmatrix}\begin{Bmatrix} x \\ y \\ 1 \end{Bmatrix} = \begin{Bmatrix} 0 \\ 0 \end{Bmatrix},$$

or using homogeneous coordinates,

$$B\ x = 0.$$

The point of intersection of these two lines is the unique non-zero solution of these equations. In homogeneous coordinates, the solutions are multiples of the following solution:

$$x_1 = \begin{vmatrix} b_1 & c_1 \\ b_2 & c_2 \end{vmatrix}, x_2 = -\begin{vmatrix} a_1 & c_1 \\ a_2 & c_2 \end{vmatrix}, x_3 = \begin{vmatrix} a_1 & b_1 \\ a_2 & b_2 \end{vmatrix}$$

if the rows of B are linearly independent (i.e., λ1 and λ2 represent distinct lines). Divide through by x3 to get Cramer's rule for the solution of a set of two linear equations in two unknowns.[27] Notice that this yields a point in the z = 1 plane only when the 2 × 2 submatrix associated with x3 has a non-zero determinant.

It is interesting to consider the case of three lines, λ1, λ2 and λ3, which yield the matrix equation,

$$\lambda_{1,2,3} : \begin{bmatrix} a_1 & b_1 & c_1 \\ a_2 & b_2 & c_2 \\ a_3 & b_3 & c_3 \end{bmatrix} \begin{Bmatrix} x \\ y \\ 1 \end{Bmatrix} = \begin{Bmatrix} 0 \\ 0 \\ 0 \end{Bmatrix}.$$

which in homogeneous form yields,

$C_x = 0.$

Clearly, this equation has the solution **x** = (0,0,0), which is not a point on the z = 1 plane E. For a solution to exist in the plane E, the coefficient matrix C must have rank 2, which means its determinant must be zero. Another way to say this is that the columns of the matrix must be linearly dependent.

Introduction to Linear Transformations

Another way to approach linear algebra is to consider linear functions on the two dimensional real plane E=$\mathbf{R}^2$. Here **R** denotes the set of real numbers. Let **x**=(x, y) be an arbitrary vector in E and consider the linear function λ: E→**R**, given by

$$\lambda : [a \quad b] \begin{Bmatrix} x \\ y \end{Bmatrix} = c,$$

or

$$A\mathbf{x} = c.$$

This transformation has the important property that if Ay=d, then

$$A(\alpha\mathbf{x} + \beta\mathbf{y}) = \alpha A\mathbf{x} + \beta A\mathbf{y} = \alpha c + \beta d.$$

This shows that the sum of vectors in E map to the sum of their images in R. This is the defining characteristic of a linear map, or linear transformation. For this case, where the image space is a real number the map is called a linear functional.

Consider the linear functional a little more carefully. Let i=(1,0) and j =(0,1) be the natural basis vectors on E, so that x=xi+yj. It is now possible to see that

$$A\mathbf{x} = A(x\mathbf{i} + y\mathbf{j}) = xA\mathbf{i} + yA\mathbf{j} = [A\mathbf{i} \quad A\mathbf{j}] \begin{Bmatrix} x \\ y \end{Bmatrix} = [a \quad b] \begin{Bmatrix} x \\ y \end{Bmatrix} = c.$$

Thus, the columns of the matrix A are the image of the basis vectors of E in R.

$$\lambda : A\mathbf{x} = [A\mathbf{v} \quad A\mathbf{w}] \begin{Bmatrix} \alpha \\ \beta \end{Bmatrix} = [d \quad e] \begin{Bmatrix} \alpha \\ \beta \end{Bmatrix} = c,$$

where Av=d and Aw=e are the images of the basis vectors v and w. This is written in matrix form as

$$[a \quad b]\begin{bmatrix} v_1 & w_1 \\ v_2 & w_2 \end{bmatrix} = [d \quad e].$$

Coordinates Relative to a Basis

This leads to the question of how to determine the coordinates of a vector **x** relative to a general basis **v** and **w** in E. Assume that we know the coordinates of the vectors, **x**, **v** and **w** in the natural basis **i**=(1,0) and **j** =(0,1). Our goal is two find the real numbers α, β, so that **x**=α**v**+β**w**, that is

$$\begin{Bmatrix} x \\ y \end{Bmatrix} = \begin{bmatrix} v_1 & w_1 \\ v_2 & w_2 \end{bmatrix}\begin{Bmatrix} \alpha \\ \beta \end{Bmatrix}.$$

To solve this equation for α, β, we compute the linear coordinate functionals σ and τ for the basis v, w, which are given by,

$$\sigma = [\sigma_1 \quad \sigma_2] = \frac{1}{v_1 w_2 - v_2 w_1}[w_2 \quad -w_1], \tau = [\tau_1 \quad \tau_2] = \frac{1}{v_1 w_2 - v_2 w_1}[-v_2 \quad v_1],$$

The functionals σ and τ compute the components of x along the basis vectors v and w, respectively, that is,

$$\sigma \mathbf{x} = \alpha, \tau \mathbf{x} = \beta,$$

which can be written in matrix form as

$$\begin{bmatrix} \sigma_1 & \sigma_2 \\ \tau_1 & \tau_2 \end{bmatrix}\begin{Bmatrix} x \\ y \end{Bmatrix} = \begin{Bmatrix} \alpha \\ \beta \end{Bmatrix}.$$

These coordinate functionals have the properties,

$$\sigma \mathbf{v} = 1, \sigma \mathbf{w} = 0, \tau \mathbf{w} = 1, \tau \mathbf{v} = 0.$$

These equations can be assembled into the single matrix equation,

$$\begin{bmatrix} \sigma_1 & \sigma_2 \\ \tau_1 & \tau_2 \end{bmatrix}\begin{bmatrix} v_1 & w_1 \\ v_2 & w_2 \end{bmatrix} = \begin{bmatrix} 1 & 0 \\ 0 & 1 \end{bmatrix}.$$

Thus, the matrix formed by the coordinate linear functionals is the inverse of the matrix formed by the basis vectors.

Inverse Image

The set of points in the plane E that map to the same image in **R** under the linear functional λ define a line in E. This line is the image of the inverse map, λ^{-1}: **R**→E. This inverse image is the set of the points **x**=(x, y) that solve the equation,

$$A\mathbf{x} = [a \quad b]\begin{Bmatrix} x \\ y \end{Bmatrix} = c.$$

Notice that a linear functional operates on known values for **x**=(x, y) to compute a value c in **R**, while the inverse image seeks the values for **x**=(x, y) that yield a specific value c.

In order to solve the equation, we first recognize that only one of the two unknowns (x,y) can be determined, so we select y to be determined, and rearrange the equation

$$by = c - ax.$$

Solve for y and obtain the inverse image as the set of points,

$$\mathbf{x}(t) = \begin{Bmatrix} 0 \\ c/b \end{Bmatrix} + t\begin{Bmatrix} 1 \\ -a/b \end{Bmatrix} = \mathbf{p} + t\mathbf{h}.$$

For convenience the free parameter x has been relabeled t.

The vector **p** defines the intersection of the line with the y-axis, known as the y-intercept. The vector **h** satisfies the homogeneous equation,

$$A\mathbf{h} = \begin{bmatrix} a & b \end{bmatrix}\begin{Bmatrix} 1 \\ -a/b \end{Bmatrix} = 0.$$

Notice that if **h** is a solution to this homogeneous equation, then t **h** is also a solution.

The set of points of a linear functional that map to zero define the kernel of the linear functional. The line can be considered to be the set of points **h** in the kernel translated by the vector **p**.

Generalizations and Related Topics

Since linear algebra is a successful theory, its methods have been developed and generalized in other parts of mathematics. In module theory, one replaces the field of scalars by a ring. The concepts of linear independence, span, basis, and dimension (which is called rank in module theory) still make sense. Nevertheless, many theorems from linear algebra become false in module theory. For instance, not all modules have a basis (those that do are called free modules), the rank of a free module is not necessarily unique, not every linearly independent subset of a module can be extended to form a basis, and not every subset of a module that spans the space contains a basis.

In multilinear algebra, one considers multivariable linear transformations, that is, mappings that are linear in each of a number of different variables. This line of inquiry naturally leads to the idea of the dual space, the vector space V^* consisting of linear maps f: V → F where F is the field of scalars. Multilinear maps T: $V^n \rightarrow F$ can be described via tensor products of elements of V^*.

If, in addition to vector addition and scalar multiplication, there is a bilinear vector product $V \times V \rightarrow V$, the vector space is called an algebra; for instance, associative algebras are algebras with an associate vector product (like the algebra of square matrices, or the algebra of polynomials).

Functional analysis mixes the methods of linear algebra with those of mathematical analysis and studies various function spaces, such as L^pspaces.

Representation theory studies the actions of algebraic objects on vector spaces by representing these objects as matrices. It is interested in all the ways that this is possible, and it does so by finding subspaces invariant under all transformations of the algebra. The concept of eigenvalues and eigenvectors is especially important.

Algebraic geometry considers the solutions of systems of polynomial equations.

There are several related topics in the field of Computer Programming that utilizes much of the techniques and theorems Linear Algebra encompasses and refers to.

REFERENCES

1. Banerjee, Sudipto; Roy, Anindya (2014), Linear Algebra and Matrix Analysis for Statistics, Texts in Statistical Science (1st ed.), Chapman and Hall/CRC, ISBN 978-1420095388
2. Strang, Gilbert (July 19, 2005), Linear Algebra and Its Applications (4th ed.), Brooks Cole, ISBN 978-0-03-010567-8
3. Weisstein, Eric. "Linear Algebra". From MathWorld--A Wolfram Web Resource. Wolfram. Retrieved 16 April 2012.
4. Vitulli, Marie. "A Brief History of Linear Algebra and Matrix Theory". Department of Mathematics. University of Oregon. Archived from the original on 2012-09-10. Retrieved 2014-07-08.
5. http://www.journals.istanbul.edu.tr/tr/index.php/oba/article/download/9103/8452
6. https://archive.org/details/linearalgebra00tevfgoog
7. Tucker, Alan (1993). "The Growing Importance of Linear Algebra in Undergraduate Mathematics". College Mathematics Journal. 24 (1): 3–9. doi:10.2307/2686426.
8. Goodlad, John I.; von stoephasius, Reneta; Klein, M. Frances (1966). "The changing school curriculum". U.S. Department of Health, Education, and Welfare: Office of Education. Retrieved 9 July 2014.
9. Dorier, Jean-Luc; Robert, Aline; Robinet, Jacqueline; Rogalsiu, Marc (2000). Dorier, Jean-Luc, ed. The Obstacle of Formalism in Linear Algebra. Springer. pp. 85–124. ISBN 978-0-7923-6539-6. Retrieved 9 July 2014.
10. Carlson, David; Johnson, Charles R.; Lay, David C.; Porter, A. Duane (1993). "The Linear Algebra Curriculum Study Group Recommendations for the First Course in Linear Algebra". The College Mathematics Journal. 24 (1): 41–46. doi:10.2307/2686430.
11. Carol S. Schumacher, Martha J. Siegel, and Paul Zorn (2015) 2015 CUPM Curriculum Guide to Majors in the Mathematical Sciences. The Mathematical Association of America. department-guidelines-recommendations/cupm
12. Peter R. Turner et al. (2015) Modeling across the Curriculum II. Report on the second SIAM-NSF Workshop, Alexandria, VA.
13. Cleve Moler, (2006) Mathworks
14. A. J. Roberts (2017) Linear Algebra Reformed for 21st-C Application.
15. Roman 2005, ch. 1, p. 27
16. Axler (2004), pp. 28–29

17. The existence of a basis is straightforward for countably generated vector spaces, and for well-ordered vector spaces, but in full generality it is logically equivalent to the axiom of choice.
18. Axler (2204), p. 33
19. Axler (2004), p. 55
20. If we restrict to integers, then only 1 and -1 have an inverse. Consequently, the inverse of an integer matrix is an integer matrix if and only if the determinant is 1 or -1.
21. P. K. Jain, Khalil Ahmad (1995). "5.1 Definitions and basic properties of inner product spaces and Hilbert spaces". Functional analysis (2nd ed.). New Age International. p. 203. ISBN 81-224-0801-X.
22. Eduard Prugovec ki (1981). "Definition 2.1". Quantum mechanics in Hilbert space (2nd ed.). Academic Press. pp. 18 ff. ISBN 0-12-566060-X.
23. Gunawardena, Jeremy. "Matrix algebra for beginners, Part I" (PDF). Harvard Medical School. Retrieved 2 May 2012.
24. Miller, Steven. "The Method of Least Squares" (PDF). Brown University. Retrieved 1 May 2013.
25. Strang, Gilbert (July 19, 2005), Linear Algebra and Its Applications (4th ed.), Brooks Cole, ISBN 978-0-03-010567-8,
26. . G. Semple and G. T. Kneebone, Algebraic Projective Geometry, Clarendon Press, London, 1952.
27. E. D. Nering, Linear Algebra and Matrix Theory, John-Wiley, New York, NY, 1963

CHAPTER 5

Mathematical Model

Chapter Contents

- **Introduction**
- **Governing Equation**
- **Defining Equation (Physics)**
- **Constitutive Equation**
- **Constraint (Mathematics)**

Mathematical models are used particularly in the natural sciences and engineering disciplines (such as physics, biology, and electrical engineering) but also in the social sciences (such as economics, sociology and political science); physicists, engineers, computer scientists, and economists use mathematical models most extensively.

INTRODUCTION

A mathematical model is a description of a system using mathematical concepts and language. The process of developing a mathematical model is termed mathematical modeling. Mathematical models are used in the natural sciences (such as physics, biology, earth science, meteorology) and engineering disciplines (such as computer science, artificial intelligence), as well as in the social sciences (such as economics, psychology, sociology, political science). Physicists, engineers, statisticians, operations research analysts, and economists use mathematical models most extensively. A model may help to explain a system and to study the effects of different components, and to make predictions about behaviour.

Elements of a Mathematical Model

Mathematical models can take many forms, including dynamical systems, statistical models, differential equations, or game theoretic models. These and other types of models can overlap, with a given model involving a variety of abstract structures. In general, mathematical models may include logical models. In many cases, the quality of a scientific field depends on how well the mathematical models developed on the theoretical side agree with results of repeatable experiments. Lack of agreement between theoretical mathematical models and experimental measurements often leads to important advances as better theories are developed.

In the physical sciences, the traditional mathematical model contains four major elements. These are

- Governing equations
- Defining equations
- Constitutive equations
- Constraints

GOVERNING EQUATION

The governing equations of a mathematical model describe how the unknown variables (i.e. the dependent variables) will change. The change of variables with respect to time may be explicit, in that a governing equation includes a derivative with respect to time, or implicit, such as when a governing equation has velocity or flux as an unknown variable.

Examples

The classic governing equations in continuum mechanics are those for the

- balance of mass;
- balance of (linear) momentum;
- balance of angular momentum;
- balance of energy; and
- balance of entropy.

For isolated systems the first four equations are the familiar conservation equations in physics. A governing equation may also take the form of a flux equation such as the diffusion equation or the heat conduction equation. In these cases the flux itself is a variable describing change of the unknown variable or property (e.g., mole concentration or internal energy or temperature).

A governing equation may also be an approximation and adaptation of the above basic equations to the situation or model in question. A governing equation may also be derived directly from experimental results and therefore be an empirical equation. A governing equation may also be a state equation, an equation describing the state of the system, and thus actually be a constitutive equation that has “stepped up the ranks” because the model in question was not meant to include a time-dependent term in the equation. This is the case for a model of a petroleum processing plant. Results from one thermodynamic equilibrium calculation are input data to the next equilibrium calculation together with some new state parameters, and so on. In this case the algorithm and sequence of input data form a chain of actions, or calculations, that describes change of states from the first state (based solely on input data) to the last state that finally comes out of the calculation sequence.

Some examples using differential equations are

- Lotka-Volterra equations are predator-prey equations
- Hele-Shaw flow
- Plate theory
  - Kirchhoff–Love plate theory or Bending of Kirchhoff-Love plates
  - Mindlin–Reissner plate theory or Bending of thick Mindlin plates or Bending of Reissner-Stein cantilever plates
- Vortex shedding
- Annular fin
- Astronautics
- Finite volume method for unsteady flow
- Acoustic theory
- Precipitation hardening
- Kelvin's circulation theorem
- Kernel function for solving integral equation of surface radiation exchanges
- Nonlinear acoustics
- Large eddy simulation
- Föppl–von Kármán equations
- Timoshenko beam theory

DEFINING EQUATION (PHYSICS)

In physics, defining equations are equations that define new quantities in terms of base quantities. This article uses the current SI system of units, not natural or characteristic units.

Description of Units and Physical Quantities

Physical quantities and units follow the same hierarchy; chosen base quantities have defined base units, from these any other quantities may be derived and have corresponding derived units.

Colour Mixing Analogy

Defining quantities is analogous to mixing colours, and could be classified a similar way, although this is not standard. Primary colours are to base quantities; as secondary (or tertiary etc.) colours are to derived quantities. Mixing colours is analogous to combining quantities using mathematical operations. But colours could be for light or paint, and analogously the system of units could be one of many forms: such as SI (now most common), CGS, Gaussian, old imperial units, a specific form of natural units or even arbitrarily defined units characteristic to the physical system in consideration (characteristic units).

The choice of a base system of quantities and units is arbitrary; but once chosen it must be adhered to throughout all analysis which follows for consistency. It makes no sense to mix up different systems of units. Choosing a system of units, one system out of the SI, CGS etc., is like choosing whether use paint or light colours.

In light of this analogy, primary definitions are base quantities with no defining equation, but defined standardized condition, "secondary" definitions are quantities defined purely in terms of base quantities, "tertiary" for quantities in terms of both base and "secondary" quantities, "quaternary" for quantities in terms of base, "secondary", and "tertiary" quantities, and so on.

Motivation

Much of physics requires definitions to be made for the equations to make sense.

Theoretical implications: Definitions are important since they can lead into new insights of a branch of physics. Two such examples occurred in classical physics. When entropy S was defined – the range of thermodynamics was greatly extended by associating chaos and disorder with a numerical quantity that could relate to energy and temperature, leading to the understanding of the second thermodynamic law and statistical mechanics.

Also the action functional (also written S) (together with generalized coordinates and momenta and the Lagrangian function), initially an alternative formulation of classical mechanics to Newton's laws, now extends the range of modern physics in general – notably quantum mechanics, particle physics, and general relativity.

Analytical convenience: They allow other equations to be written more compactly and so allow easier mathematical manipulation; by including a parameter in a definition, occurrences of the parameter can be absorbed into the substituted quantity and removed from the equation.

Example

As an example consider Ampère's circuital law (with Maxwell's correction) in integral form for an arbitrary current carrying conductor in a vacuum (so zero magnetization due medium, i.e. $\mathbf{M} = \mathbf{0}$):

$$\oint_S \mathbf{B} \cdot \mathbf{dl} = \mu_0 \oint_S \left(\mathbf{J} + \epsilon_0 \frac{\partial \mathbf{E}}{\partial t}\right) \cdot \mathbf{dA}$$

using the constitutive definition

$$\mathbf{B} = \mu_0 \mathbf{H},$$

and the current density definition

$$I = \oint_S \mathbf{J} \cdot \mathbf{dA},$$

similarly for the displacement current density

$\mathbf{J}_\mathrm{d} = \epsilon_0 \dfrac{\partial \mathbf{E}}{\partial t}$ leading to the displacement current $I_d = \oint_S \mathbf{J}_\mathrm{d} \cdot \mathbf{dA},$

we have

$$\oint_S \mathbf{B} \cdot \mathbf{dl} = \mu_0 \oint_S \mathbf{J} \cdot \mathbf{dA} + \mu_0 \oint_S \mathbf{J}_\mathrm{d} \cdot \mathbf{dA},$$

$$\oint_S \mathbf{H} \cdot \mathbf{dl} = I + I_d,$$

which is simpler to write, even if the equation is the same.

Ease of comparison: They allow comparisons of measurements to be made when they might appear ambiguous and unclear otherwise.

Example

A basic example is mass density. It is not clear how compare how much matter constitutes a variety of substances given only their masses or only their volumes. Given both for each substance, the mass m per unit volume V, or mass density ρ provides a meaningful comparison between the substances, since for each, a fixed amount of volume will correspond to an amount of mass depending on the substance. To illustrate this; if two substances A and B have masses m_A and m_B respectively, occupying volumes V_A and V_B respectively, using the definition of mass density gives:

$\rho_A = m_A / V_A$, $\rho_B = m_B / V_B$

following this can be seen that:

- if $m_A > m_B$ or $m_A < m_B$ and $V_A = V_B$, then $\rho_A > \rho_B$ or $\rho_A < \rho_B$,
- if $m_A = m_B$ and $V_A > V_B$ or $V_A < V_B$, then $\rho_A < \rho_B$ or $\rho_A > \rho_B$,
- if $\rho_A = \rho_B$, then $m_A / V_A = m_B / V_B$ so $m_A / m_B = V_A / V_B$, demonstrating that if $m_A > m_B$ or $m_A < m_B$, then $V_A > V_B$ or $V_A < V_B$.

Making such comparisons without using mathematics logically in this way would not be as systematic.

Construction of Defining Equations

Scope of Definitions

Defining equations are normally formulated in terms of elementary algebra and calculus, vector algebra and calculus, or for the most general applications tensor algebra and calculus, depending on the level of study and presentation, complexity of topic and scope of applicability. Functions may be incorporated into a definition, in for calculus this is necessary. Quantities may also be complex-valued for theoretical advantage, but for a physical measurement the real part is relevant, the imaginary part can be discarded. For more advanced treatments the equation may have to be written in an equivalent but alternative form using other defining equations for the definition to be useful. Often definitions can start from elementary algebra, then modify to vectors, then in the limiting cases calculus may be used. The various levels of maths used typically follows this pattern.

Typically definitions are explicit, meaning the defining quantity is the subject of the equation, but sometimes the equation is not written explicitly – although the defining quantity can be solved for to make the equation explicit. For vector equations, sometimes the defining quantity is in a cross or dot product and cannot be solved for explicitly as a vector, but the components can.

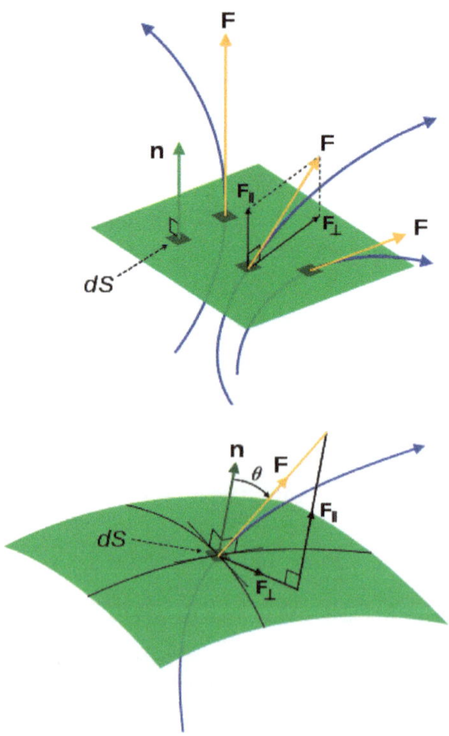

Flux **F** through a surface, d**S** is the differential vector area element, n is the unit normal to the surface. For the physical examples here, current density **J** or magnetic field **B** would be **F** in the diagram.

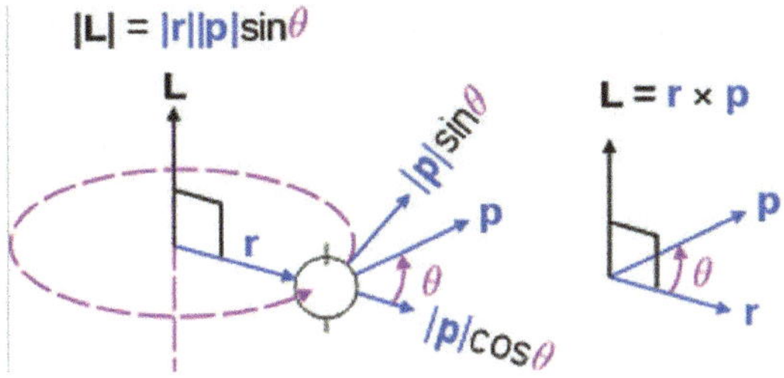

Angular momentum; scalar and vector components.

Examples

Electric current density is an example spanning all of these methods, Angular momentum is an example which doesn't require calculus. See the classical mechanics section below for nomenclature and diagrams to the right.

Elementary algebra

Operations are simply multiplication and division. Equations may be written in a product or quotient form, both of course equivalent.

	Angular momentum	Electric current density
Quotient form	$p = \frac{L}{r}$	$J = \frac{I}{A}$
Product form	$L = pr$	$I = JA$

Vector algebra

There is no way to divide a vector by a vector, so there are no product or quotient forms.

	Angular momentum	Electric current density
Quotient form	N/A	$\mathbf{J} \cdot \hat{\mathbf{n}} = \frac{I}{A}$
Product form	Starting from $L = pr,$	$\mathbf{J} \cdot \hat{\mathbf{n}}A = I,$ $\mathbf{J} \cdot \mathbf{A} = I,$

since $\mathbf{L} = \mathbf{0}$ when $\mathbf{p}$ and $\mathbf{r}$ are parallel or antiparallel, and is a maximum when perpendicular, so that the only component of $\mathbf{p}$ which contributes to $\mathbf{L}$ is the tangential $|\mathbf{p}| \sin \theta$, the magnitude of angular momentum L should be re-written as

$$L = pr\sin\theta.$$

Since r, p and L form a right-hand triad, this leads to the vector form

$$\mathbf{L} = \mathbf{r} \times \mathbf{p}.$$

Elementary Calculus

The arithmetic operations are modified to the limiting cases of differentiation and integration. Equations can be expressed in these equivalent and alternative ways.

	Current density
Differential form	$J = \lim_{A\to 0} \frac{I}{A} = \frac{\mathrm{d}I}{\mathrm{d}A}$
Integral form	$I = \lim_{A_i \to 0} \sum_i JA_i = \int_S J\mathrm{d}A$

where dA means an differential area element .

Alternatively for integral form

$$\mathrm{d}I = J\mathrm{d}A,$$

$$I = \int_S J\mathrm{d}A.$$

Vector calculus

	Current density
Differential form	$\mathbf{J}\cdot\hat{\mathbf{n}} = \frac{\mathrm{d}I}{\mathrm{d}A}$
Integral form	$I = \int_S \mathbf{J}\cdot\mathrm{d}\mathbf{A}$

where d**A** = **n**dA is the differential vector area.

Tensor analysis

Vectors are rank-1 tensors. The formulae below are no more than the vector equations in the language of tensors.

	Angular momentum	Electric current density
Differential form	N/A	$J_i n_i = \frac{\mathrm{d}I}{\mathrm{d}A}$
Product/integral form	Starting from $\mathbf{L} = \mathbf{r} \times \mathbf{p}$ the components are L_i, r_j, p_i, where i, j, k are each dummy indices each taking values 1, 2, 3, using the identity from tensor analysis $\mathbf{a} = \mathbf{b} \times \mathbf{c}, \quad a_i = \epsilon_{ijk} b_j c_k,$ where ε_{ijk} is the permutation/Levi-Cita tensor, leads to $L_i = \epsilon_{ijk} r_j p_k.$	Using the Einstein summation convention, $J_i n_i \mathrm{d}A = \mathrm{d}I$ $\int_S J_i \mathrm{d}A_i = I$

Multiple Choice Definitions

Sometimes there is still freedom within the chosen units system, to define one or more quantities in more than one way. The situation splits into two cases:

Mutually exclusive definitions: There are a number of possible choices for a quantity to be defined in terms of others, but only one can be used and not the others. Choosing more than one of the exclusive equations for a definition leads to a contradiction – one equation might demand a quantity X to be defined in one way using another quantity Y, while another equation requires the reverse, Y be defined using X, but then another equation might falsify the use of both X and Y, and so on. The mutual disagreement makes it impossible to say which equation defines what quantity.

Equivalent definitions: Defining equations which are equivalent and self-consistent with other equations and laws within the physical theory, simply written in different ways.

There are two possibilities for each case:

One defining equation – one defined quantity: A defining equation is used to define a single quantity in terms of a number of others.

One defining equation – a number of defined quantities: A defining equation is used to define a number of quantities in terms of a number of others. A single defining equation shouldn't contain one quantity defining all other quantities in the same equation, otherwise contradictions arise again. There is no definition of the defined quantities separately since they are defined by a single quantity in a single equation. Furthermore, the defined quantities may have already been defined before, so if another quantity defines these in the same equation, there is a clash between definitions.

Contradictions can be avoided by defining quantities successively; the order in which quantities are defined must be accounted for. Examples spanning these instances occur in electromagnetism, and are given below.

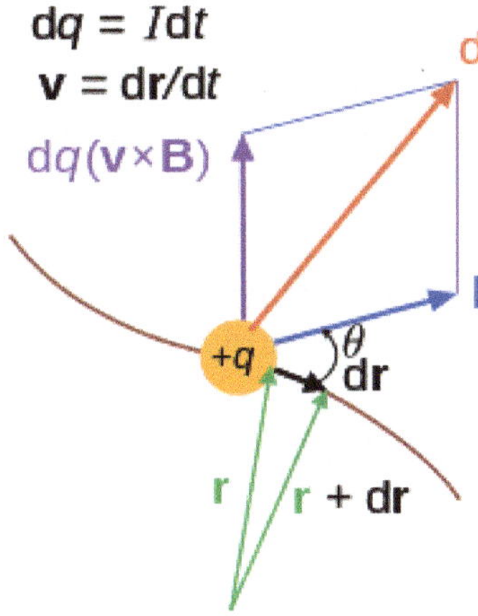

Differential magnetic force **dF** due to a small charge element dq constituting an electric current I (conventional current is used). The force must be line-integrated along the path of current flow, with respect to the vector line element **dr**.

Examples

Mutually Exclusive Definitions:

The magnetic induction field **B** can be defined in terms of electric charge q or current I, and the Lorentz force (magnetic term) **F** experienced by the charge carriers due to the field,

$$\begin{aligned}\mathbf{F} &= q(\mathbf{v}\times\mathbf{B})\\ &= \left(\int I\mathrm{d}t\right)\left(\frac{\mathrm{d}\mathbf{r}}{\mathrm{d}t}\times\mathbf{B}\right)\\ &= \left(\int I\mathrm{d}t\frac{\mathrm{d}\mathbf{r}}{\mathrm{d}t}\right)\times\mathbf{B}\\ &= I\left(\int \mathrm{d}\mathbf{r}\right)\times\mathbf{B}\\ &= I(\mathbf{l}\times\mathbf{B}),\end{aligned}$$

where $\mathbf{l} = \int \mathrm{d}\mathbf{r}$ is the change in position traversed by the charge carriers (assuming current is independent of position, if not so a line integral must be done along the path of current) or in terms of the magnetic flux Φ_R through a surface S, where the area is used as a scalar A and vector: $\mathbf{A} = A\hat{\mathbf{n}}$ and $\hat{\mathbf{n}}$ is a unit normal to A, is a unit normal to A, either in differential form

$$\mathbf{B}\cdot\hat{\mathbf{n}} = \frac{\mathrm{d}\Phi_B}{\mathrm{d}A},$$

or integral form,

$$\mathbf{B}\cdot\hat{\mathbf{n}}\mathrm{d}A = \mathrm{d}\Phi_B,$$

$$\Phi_B = \int_S \mathbf{B}\cdot\mathrm{d}\mathbf{A}.$$

However, only one of the above equations can be used to define **B** for the following reason, given that **A**, **r**, **v**, and **F** have been defined elsewhere unambiguously (most likely mechanics and Euclidean geometry).

If the force equation defines **B**, where q or I have been previously defined, then the flux equation defines Φ_B, since **B** has been previously defined unambiguously. If the flux equation defines **B**, where Φ_B, the force equation may be a defining equation for I or q. Notice the contradiction when **B** both equations define **B** simultaneously and when **B** is not a base quantity; the force equation demands that q or I be defined elsewhere while at the same time the flux equation demands that q or I be defined by the force equation, similarly the force equation requires Φ_B to be defined by the flux equation, at the same time the flux equation demands that Φ_B is defined elsewhere. For both equations to be used as definitions simultaneously, **B** must be a base quantity so that **F** and Φ_B can be defined to stem from **B** unambiguously.

Equivalent Definitions:

Another example is inductance L which has two equivalent equations to use as a definition.

In terms of I and Φ_B, the inductance is given by

$$L = N\frac{d\Phi_B}{dI},$$

in terms of *I* and induced emf *V*

$$V = -L\frac{dI}{dt}.$$

These two are equivalent by Faraday's law of induction:

$$V = -N\frac{d\Phi_B}{dt},$$

$$Vdt = -Nd\Phi_B,$$

substituting into the first definition for *L*

$$L = -V\frac{dt}{dI}$$

$$V = -L\frac{dI}{dt}$$

and so they are not mutually exclusive.

One defining equation – a number of defined quantities

Notice that L cannot define I and Φ_B simultaneously - this makes no sense. I, Φ_B and V have most likely all been defined before as (Φ_B given above in flux equation);

$$V = \frac{dW}{dq}, \quad I = \frac{dq}{dt},$$

where W = work done on charge q. Furthermore, there is no definition of either I or Φ_B separately – because L is defining them in the same equation.

However, using the Lorentz force for the electromagnetic field:

$$\mathbf{F} = q\left[\mathbf{E} + (\mathbf{v} \times \mathbf{B})\right],$$

as a single defining equation for the electric field **E** and magnetic field **B** is allowed, since **E** and **B** are not only defined by one variable, but three; force **F**, velocity **v** and charge q. This is consistent with isolated definitions of **E** and **B** since **E** is defined using **F** and q:

$$\mathbf{E} = \mathbf{F}/q.$$

and **B** defined by **F**, v, and *q*, as given above.

Limitations of Definitions

Definitions vs. functions: Defining quantities can vary as a function of parameters other than those in the definition. A defining equation only defines how calculate the defined quantity, it cannot describe how the quantity varies as a function of other parameters since the function would vary from one application to another.

Examples

Mass density ρ is defined using mass m and volume V by but can vary as a function of temperature T and pressure p, $\rho = \rho(p, T)$

The angular frequency ω of wave propagation is defined using the frequency (or equivalently time period T) of the oscillation, as a function of wavenumber k, $\omega = \omega(k)$. This is the dispersion relation for wave propagation.

The coefficient of restitution for an object colliding is defined using the speeds of separation and approach with respect to the collision point, but depends on the nature of the surfaces in question.

Definitions vs. theorems: There is a very important difference between defining equations and general or derived results, theorems or laws. Defining equations do ***not*** find out ***any*** information about a physical system, they simply re-state one measurement in terms of others. Results, theorems, and laws, on the other hand do provide meaningful information, if only a little, since they represent a calculation for a quantity given other properties of the system, and describe how the system behaves as variables are changed.

Examples

An example was given above for Ampere's law. Another is the conservation of momentum for N_1 initial particles having initial momenta $\mathbf{p}_i$ where i = 1, 2 ... N_1, and N_2 final particles having final momenta $\mathbf{p}_i$ (some particles may explode or adhere) where j = 1, 2 ... N_2, the equation of conservation reads:

$$\sum_i^{N_1} \mathbf{p}_\mathrm{i} = \sum_j^{N_2} \mathbf{p}_\mathrm{j}$$

Using the definition of momentum in terms of velocity:

$$\mathbf{p} = m\mathbf{v}$$

so that for each particle:

$$\mathbf{p}_\mathrm{i} = m_i \mathbf{v}_\mathrm{i} \text{ and } \mathbf{p}_\mathrm{j} = m_j \mathbf{v}_\mathrm{j}$$

the conservation equation can be written as

$$\sum_i^{N_1} m_i \mathbf{v}_\mathrm{i} = \sum_j^{N_2} m_i \mathbf{v}_\mathrm{i}.$$

It is identical to the previous version. No information is lost or gained by changing quantities when definitions are substituted, but the equation itself does give information about the system.

One-Off Definitions

Some equations, typically results from a derivation, include useful quantities which serve as a one-off definition within its scope of application.

Examples

In special relativity, relativistic mass has support and detraction by physicists. It is defined as:

$m = \gamma m_0$

where m_0 is the rest mass of the object and γ is the Lorentz factor. This makes some quantities such as momentum **p** and energy E of a massive object in motion easy to obtain from other equations simply by using relativistic mass:

$\mathbf{p} = m\mathbf{v} \rightarrow \mathbf{p} = \gamma m_0 \mathbf{v}$

$E = mc^2 \rightarrow E = \gamma m_0 c^2$

However, this does not always apply, for instance the kinetic energy T and force **F** of the same object is not given by:

$T = \frac{m}{2}\mathbf{v}\cdot\mathbf{v} \nrightarrow T = \frac{\gamma m_0}{2}\mathbf{v}\cdot\mathbf{v}$

$\mathbf{F} = m\mathbf{a} \nrightarrow \mathbf{F} = \gamma m_0 \mathbf{a}$

The Lorentz factor has a deeper significance and origin, and is used in terms of proper time and coordinate time with four-vectors. The correct equations above are consequence of the applying definitions in the correct order.

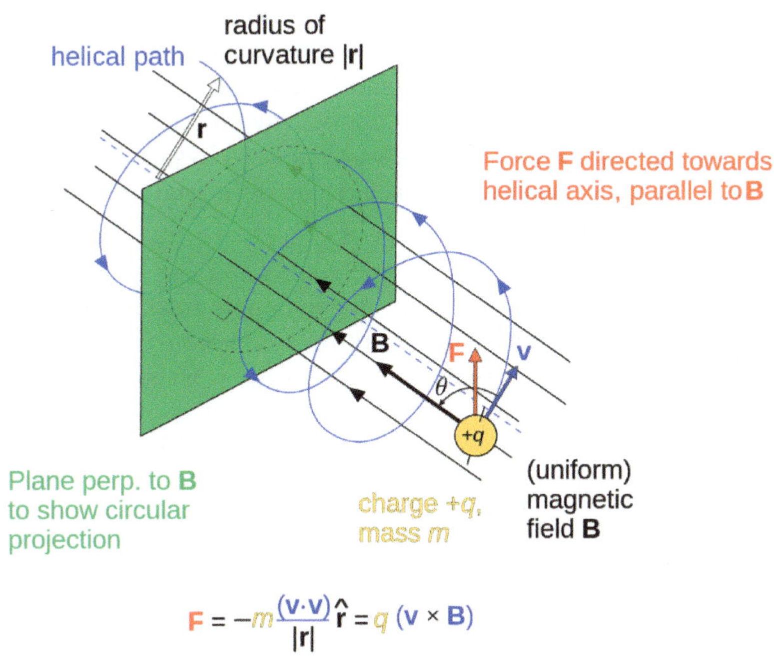

Magnetic field deflecting a charged particle, pseudo-defining magnetic rigidity for the particle.

In electromagnetism, a charged particle (of mass m and charge q) in a uniform magnetic field **B** is deflected by the field in a circular helical arc at velocity **v** and radius of curvature **r**, where the helical trajectory inclined at an angle θ to **B**. The magnetic force is the centripetal force, so the force **F** acting on the particle is;

$$\mathbf{F} = -\frac{m(\mathbf{v}\cdot\mathbf{v})\,\hat{\mathbf{r}}}{|\mathbf{r}|} = q(\mathbf{v}\times\mathbf{B}),$$

reducing to scalar form and solving for $|\mathbf{B}||\mathbf{r}|$:

$$\frac{m|\mathbf{v}|^2}{|\mathbf{r}|} = q\,|\mathbf{v}|\,|\mathbf{B}|\sin\theta,$$

$$\frac{m\,|\mathbf{v}|}{|\mathbf{r}|} = q\,|\mathbf{B}|\sin\theta,$$

$$|\mathbf{B}|\,|\mathbf{r}| = \frac{m\,|\mathbf{v}|}{q\sin\theta},$$

serves as the definition for the **magnetic rigidity** of the particle. Since this depends on the mass and charge of the particle, it is useful for determining the extent a particle deflects in a **B** field, which occurs experimentally in mass spectrometry and particle detectors.

CONSTITUTIVE EQUATION

In physics and engineering, a constitutive equation or constitutive relation is a relation between two physical quantities (especially kinetic quantities as related to kinematic quantities) that is specific to a material or substance, and approximates the response of that material to external stimuli, usually as applied fields or forces. They are combined with other equations governing physical laws to solve physical problems; for example in fluid mechanics the flow of a fluid in a pipe, in solid state physics the response of a crystal to an electric field, or in structural analysis, the connection between applied stresses or forces to strains or deformations.

Some constitutive equations are simply phenomenological; others are derived from first principles. A common approximate constitutive equation frequently is expressed as a simple proportionality using a parameter taken to be a property of the material, such as electrical conductivity or a spring constant. However, it is often necessary to account for the directional dependence of the material, and the scalar parameter is generalized to a tensor. Constitutive relations are also modified to account for the rate of response of materials and their non-linear behavior. See the article Linear response function.

Mechanical Properties of Matter

The first constitutive equation (constitutive law) was developed by Robert Hooke and is known as Hooke's law. It deals with the case of linear elastic materials. Following this discovery, this type of equation, often called a "stress-strain relation" in this example, but also called a "constitutive assumption" or an "equation of state" was commonly used. Walter Noll advanced the use of constitutive equations, clarifying their classification and the role of invariance requirements, constraints, and definitions of terms like "material", "isotropic", "aeolotropic", etc. The class of "constitutive relations" of the form stress rate = f (velocity gradient, stress, density) was the subject of Walter Noll's dissertation in 1954 under Clifford Truesdell.

In modern condensed matter physics, the constitutive equation plays a major role. See Linear constitutive equations and Nonlinear correlation functions.

Definitions

Quantity (common name/s)	(Common) symbol/s	Defining equation	SI units	**Dimension**
General stress, Pressure	P, σ	$\sigma = F/A$ F may be any perpendicular force applied to area A	Pa = N m^{-2}	[M] $[T]^{-2}[L]^{-1}$
General strain	ε	$\varepsilon = \Delta D/D$ D = dimension (length, area, volume) ΔK = change in material	dimensionless	dimensionless
General elastic modulus	E_{mod}	$E_{mod} = \sigma/\varepsilon$	Pa = N m^{-2}	[M] $[T]^{-2}$ $[L]^{-1}$
Young's modulus	E, Y	$Y = \sigma/(\Delta L/L)$	Pa = N m^{-2}	[M] [T] $^{-2}[L]^{-1}$
Shear modulus	G	$G = \Delta x/L$	Pa = N m^{-2}	[M] $[T]^{-2}[L]^{-1}$
Bulk modulus	K, B	$B = P/(\Delta V/V)$	Pa = N m^{-2}	[M] $[T]^{-2}[L]^{-1}$
Compressibility	C	$C = 1/B$	Pa^{-1} = m^2 N^{-1}	[L] $[T]^2[M]^{-1}$

Deformation of Solids

Friction

Friction is a complicated phenomenon. Macroscopically the friction force F between the interface of two materials can be modelled as proportional to the reaction force R at a point of contact between two interfaces, through a dimensionless coefficient of friction μ_f which depends on the pair of materials:

$$F = \mu_f R.$$

This can be applied to static friction (friction preventing two stationary objects from slipping on their own), kinetic friction (friction between two objects scraping/sliding past each other), or rolling (frictional force which prevents slipping but causes a torque to exert on a round object). Surprisingly, the friction force does not depend on the surface area of common contact.

Stress and Strain

The stress-strain constitutive relation for linear materials is commonly known as Hooke's law. In its simplest form, the law defines the spring constant (or elasticity constant) k in a scalar equation, stating the tensile/compressive force is proportional to the extended (or contracted) displacement x:

$$F_i = -kx_i$$

meaning the material responds linearly. Equivalently, in terms of the stress σ, Young's modulus E, and strain ε (dimensionless):

$$\sigma = E\varepsilon$$

In general, forces which deform solids can be normal to a surface of the material (normal forces), or tangential (shear forces), this can be described mathematically using the stress tensor:

$$\sigma_{ij} = C_{ijkl}\,\varepsilon_{kl} \rightleftharpoons \varepsilon_{ij} = S_{ijkl}\,\sigma_{kl}$$

where C is the elasticity tensor and S is the compliance tensor

Solid-State Deformations

Several classes of deformations in elastic materials are the following:

- Elastic: The material recovers its initial shape after deformation.
- Anelastic: if the material is close to elastic, but the applied force induces additional time-dependent resistive forces (i.e. depend on rate of change of extension/compression, in addition to the extension/compression). Metals and ceramics have this characteristic, but it is usually negligible, although not so much when heating due to friction occurs (such as vibrations or shear stresses in machines).
- Viscoelastic: If the time-dependent resistive contributions are large, and cannot be neglected. Rubbers and plastics have this property, and certainly do not satisfy Hooke's law. In fact, elastic hysteresis occurs.
- Plastic: The applied force induces non-recoverable deformations in the material when the stress (or elastic strain) reaches a critical magnitude, called the yield point.
- Hyperelastic: The applied force induces displacements in the material following a strain energy density function.

Collisions

The relative speed of separation $v_{separation}$ of an object A after a collision with another object B is related to the relative speed of approach $v_{approach}$ by the coefficient of restitution, defined by Newton's experimental impact law:

$$e = \frac{|\mathbf{v}|_{\text{separation}}}{|\mathbf{v}|_{\text{approach}}}$$

which depends the materials A and B are made from, since the collision involves interactions at the surfaces of A and B. Usually $0 \leq e \leq 1$, in which e = 1 for completely elastic collisions, and e = 0 for completely inelastic collisions. It's possible for $e \geq 1$ to occur – for superelastic (or explosive) collisions.

Deformation of Fluids

The drag equation gives the drag force D on an object of cross-section area A moving through a fluid of density ρ at velocity v (relative to the fluid)

$$\tau = \mu \frac{\partial u}{\partial y},$$

where the drag coefficient (dimensionless) c_d depends on the geometry of the object and the drag forces at the interface between the fluid and object.

For a Newtonian fluid of viscosity μ, the shear stress τ is linearly related to the strain rate (transverse flow velocity gradient) $\partial u/\partial y$ (units s^{-1}). In a uniform shear flow:

with u(y) the variation of the flow velocity u in the cross-flow (transverse) direction y. In general, for a Newtonian fluid, the relationship between the elements τij of the shear stress tensor and the deformation of the fluid is given by

$$\tau_{ij} = 2\mu\left(e_{ij} - \frac{1}{3}\Delta\delta_{ij}\right) \quad \text{with} \quad e_{ij} = \frac{1}{2}\left(\frac{\partial v_i}{\partial x_j} + \frac{\partial v_j}{\partial x_i}\right)$$

$$\text{and} \quad \Delta = \sum_k e_{kk} = \text{div } \mathbf{v},$$

where v_i are the components of the flow velocity vector in the corresponding x_i coordinate directions, e_{ij} are the components of the strain rate tensor, Δ is the volumetric strain rate (or dilatation rate) and δ_{ij} is the Kronecker delta.

The ideal gas law is a constitutive relation in the sense the pressure p and volume V are related to the temperature T, via the number of moles n of gas:

$$pV = nRT$$

where R is the gas constant (J K−1 mol−1).

Electromagnetism

Constitutive Equations in Electromagnetism and Related Areas

In both classical and quantum physics, the precise dynamics of a system form a set of coupled differential equations, which are almost always too complicated to be solved exactly, even at the level of statistical mechanics. In the context of electromagnetism, this remark applies to not only the dynamics of free charges and currents (which enter Maxwell's equations directly), but also the dynamics of bound charges and currents (which enter Maxwell's equations through the constitutive relations). As a result, various approximation schemes are typically used.

For example, in real materials, complex transport equations must be solved to determine the time and spatial response of charges, for example, the Boltzmann equation or the Fokker–Planck equation or the Navier-Stokes equations. For example, see magnetohydrodynamics, fluid dynamics, electrohydrodynamics, superconductivity, plasma modeling. An entire physical apparatus for dealing with these matters has developed. See for example, linear response theory, Green–Kubo relations and Green's function (many-body theory).

These complex theories provide detailed formulas for the constitutive relations describing the electrical response of various materials, such as permittivities, permeabilities, conductivities and so forth.

It is necessary to specify the relations between displacement field **D** and **E**, and the magnetic H-field **H** and **B**, before doing calculations in electromagnetism (i.e. applying Maxwell's macroscopic equations). These equations specify the response of bound charge and current to the applied fields and are called constitutive relations.

Determining the constitutive relationship between the auxiliary fields **D** and **H** and the **E** and **B** fields starts with the definition of the auxiliary fields themselves:

$$\mathbf{D}(\mathbf{r},t) = \varepsilon_0 \mathbf{E}(\mathbf{r},t) + \mathbf{P}(\mathbf{r},t)$$

$$\mathbf{H}(\mathbf{r},t) = \frac{1}{\mu_0}\mathbf{B}(\mathbf{r},t) - \mathbf{M}(\mathbf{r},t),$$

where **P** is the polarization field and **M** is the magnetization field which are defined in terms of microscopic bound charges and bound current respectively. Before getting to how to calculate **M** and **P** it is useful to examine the following special cases.

Without Magnetic or Dielectric Materials

In the absence of magnetic or dielectric materials, the constitutive relations are simple:

$$\mathbf{D} = \varepsilon_0 \mathbf{E}, \quad \mathbf{H} = \mathbf{B}/\mu_0$$

where ε_0 and μ_0 are two universal constants, called the permittivity of free space and permeability of free space, respectively.

Isotropic Linear Materials

In an (isotropic) linear material, where **P** is proportional to **E**, and **M** is proportional to **B**, the constitutive relations are also straightforward. In terms of the polarization **P** and the magnetization **M** they are:

$$\mathbf{P} = \varepsilon_0 \chi_e \mathbf{E}, \quad \mathbf{M} = \chi_m \mathbf{H},$$

where χ_e and χ_m are the electric and magnetic susceptibilities of a given material respectively. In terms of **D** and **H** the constitutive relations are:

$$\mathbf{D} = \varepsilon \mathbf{E}, \quad \mathbf{H} = \mathbf{B}/\mu,$$

where ε and μ are constants (which depend on the material), called the permittivity and permeability, respectively, of the material. These are related to the susceptibilities by:

$$\varepsilon/\varepsilon_0 = \varepsilon_r = (\chi_e + 1), \quad \mu/\mu_0 = \mu_r = (\chi_m + 1)$$

General Case

For real-world materials, the constitutive relations are not linear, except approximately. Calculating the constitutive relations from first principles involves determining how **P** and **M** are created from a given **E** and **B**. These relations may be empirical (based directly upon measurements), or theoretical (based upon statistical mechanics, transport theory or other tools of condensed matter physics). The detail employed may be macroscopic or microscopic, depending upon the level necessary to the problem under scrutiny.

In general, the constitutive relations can usually still be written:

$$\mathbf{D} = \varepsilon\mathbf{E}, \quad \mathbf{H} = \mu^{-1}\mathbf{B}$$

but ε and μ are not, in general, simple constants, but rather functions of **E**, **B**, position and time, and tensorial in nature. Examples are:

- Dispersion and absorption where ε and μ are functions of frequency. (Causality does not permit materials to be nondispersive; see, for example, Kramers–Kronig relations). Neither do the fields need to be in phase which leads to ε and μ being complex. This also leads to absorption.
- Nonlinearity where ε and μ are functions of **E** and **B**.
- Anisotropy (such as birefringence or dichroism) which occurs when ε and μ are second-rank tensors,

$$D_i = \sum_j \varepsilon_{ij} E_j \quad B_i = \sum_j \mu_{ij} H_j.$$

- Dependence of **P** and **M** on **E** and **B** at other locations and times. This could be due to spatial inhomogeneity; for example in a domained structure, heterostructure or a liquid crystal, or most commonly in the situation where there are simply multiple materials occupying different regions of space. Or it could be due to a time varying medium or due to hysteresis. In such cases **P** and **M** can be calculated as:

$$\mathbf{P}(\mathbf{r},t) = \varepsilon_0 \int \mathrm{d}^3\mathbf{r}'\mathrm{d}t' \; \hat{\chi}_e(\mathbf{r},\mathbf{r}',t,t';\mathbf{E})\,\mathbf{E}(\mathbf{r}',t')$$

$$\mathbf{M}(\mathbf{r},t) = \frac{1}{\mu_0} \int \mathrm{d}^3\mathbf{r}'\mathrm{d}t' \; \hat{\chi}_m(\mathbf{r},\mathbf{r}',t,t';\mathbf{B})\,\mathbf{B}(\mathbf{r}',t'),$$

in which the permittivity and permeability functions are replaced by integrals over the more general electric and magnetic susceptibilities. In homogenous materials, dependence on other locations is known as spatial dispersion.

As a variation of these examples, in general materials are bianisotropic where **D** and **B** depend on both **E** and **H**, through the additional coupling constants ξ and ζ:

$$\mathbf{D} = \varepsilon\mathbf{E} + \xi\mathbf{H}, \quad \mathbf{B} = \mu\mathbf{H} + \zeta\mathbf{E}.$$

In practice, some materials properties have a negligible impact in particular circumstances, permitting neglect of small effects. For example: optical nonlinearities can be neglected for low field strengths; material dispersion is unimportant when frequency is limited to a narrow bandwidth; material absorption can be neglected for wavelengths for which a material is transparent; and metals with finite conductivity often are approximated at microwave or longer wavelengths as perfect metals with infinite conductivity (forming hard barriers with zero skin depth of field penetration).

Some man-made materials such as metamaterials and photonic crystals are designed to have customized permittivity and permeability.

Calculation of Constitutive Relations

The theoretical calculation of a material's constitutive equations is a common, important, and sometimes difficult task in theoretical condensed-matter physics and materials science. In general, the constitutive equations are theoretically determined by calculating how a molecule responds to the local fields through the Lorentz force. Other forces may need to be modeled as well such as lattice vibrations in crystals or bond forces. Including all of the forces leads to changes in the molecule which are used to calculate **P** and **M** as a function of the local fields. The local fields differ from the applied fields due to the fields produced by the polarization and magnetization of nearby material; an effect which also needs to be modeled. Further, real materials are not continuous media; the local fields of real materials vary wildly on the atomic scale. The fields need to be averaged over a suitable volume to form a continuum approximation.

These continuum approximations often require some type of quantum mechanical analysis such as quantum field theory as applied to condensed matter physics. See, for example, density functional theory, Green–Kubo relations and Green's function.

A different set of homogenization methods (evolving from a tradition in treating materials such as conglomerates and laminates) are based upon approximation of an inhomogeneous material by a homogeneous effective medium (valid for excitations with wavelengths much larger than the scale of the inhomogeneity).

The theoretical modeling of the continuum-approximation properties of many real materials often rely upon experimental measurement as well. For example, ε of an insulator at low frequencies can be measured by making it into a parallel-plate capacitor, and ε at optical-light frequencies is often measured by ellipsometry.

Thermoelectric and Electromagnetic Properties of Matter

These constitutive equations are often used in crystallography, a field of solid-state physics.

Electromagnetic properties of solids

Property/effect	Stimuli/response parameters of system	Constitutive tensor of system	Equation
Hall effect	E = electric field strength (N C $^{-1}$) J = electric current density (A m^{-2}) H = magnetic field intensity (A m^{-1})	ρ = electrical resistivity (Ω m)	$E_k = \rho_{kij} J_i H_j$
Direct Piezoelectric Effect	σ = Stress (Pa) P = (dielectric) polarization (C m^{-2})	d = direct piezoelectric coefficient (K^{-1})	$P_i = d_{ijk} \sigma_{jk}$
Converse Piezoelectric Effect	ε = Strain (dimensionless) E = electric field strength (N C $^{-1}$)	d = direct piezoelectric coefficient (K^{-1})	$\varepsilon_{ij} = d_{ijk} E_k$
Piezomagnetic effect	σ = Stress (Pa) M = magnetization (A m^{-1})	q = piezomagnetic coefficient (K^{-1})	$M_i = q_{ijk} \sigma_{jk}$

Thermoelectric properties of solids

Property/effect	Stimuli/response parameters of system	Constitutive tensor of system	Equation
Pyroelectricity	P = (dielectric) polarization (C m^{-2}) T = temperature (K)	p = pyroelectric coefficient (C m^{-2} K^{-1})	$\Delta P_j = p_j \Delta T$
Electrocaloric effect	S = entropy (J K^{-1}) E = electric field strength (N C $^{-1}$)	p = pyroelectric coefficient (C m^{-2} K^{-1})	$\Delta S = p_i \Delta E_i$
Seebeck effect	E = electric field strength (N C $^{-1}$ = V m^{-1}) T = temperature (K) x = displacement (m)	β = thermopower (V K^{-1})	$E_i = -\beta_{ij} \dfrac{\partial T}{\partial x_j}$
Peltier effect	E = electric field strength (N C $^{-1}$) J = electric current density (A m^{-2}) q = heat flux (W m^{-2})	Π = Peltier coefficient (W A $^{-1}$)	$q_j = \Pi_{ji} J_i$

Photonics

The (absolute) refractive index of a medium n (dimensionless) is an inherently important property of geometric and physicaloptics defined as the ratio of the luminal speed in vacuum c_0 to that in the medium c:

$$n = \frac{c_0}{c} = \sqrt{\frac{\varepsilon\mu}{\varepsilon_0\mu_0}} = \sqrt{\varepsilon_r\mu_r}$$

where ε is the permittivity and ε_r the relative permittivity of the medium, likewise μ is the permeability and μ_r are the relative permmeability of the medium. The vacuum permittivity is ε_0 and vacuum permeability is μ_0. In general, n (also ε_r) are complex numbers.

The relative refractive index is defined as the ratio of the two refractive indices. Absolute is for on material, relative applies to every possible pair of interfaces;

$$n_{AB} = \frac{n_A}{n_B}$$

Speed of light in matter

As a consequence of the definition, the speed of light in matter is

$$c = 1/\sqrt{\varepsilon\mu}$$

for special case of vacuum; $\varepsilon = \varepsilon_0$ and $\mu = \mu_0$,

$$c_0 = 1/\sqrt{\varepsilon_0\mu_0}$$

Piezooptic Effect

The piezooptic effect relates the stresses in solids σ to the dielectric impermeability a, which are coupled by a fourth-rank tensor called the piezooptic coefficient Π (units K^{-1}):

$$a_{ij} = \Pi_{ijpq}\sigma_{pq}$$

Transport Phenomena

Definitions

Definitions (thermal properties of matter)

Quantity (Common Name/s)	(Common) Symbol/s	Defining Equation	SI Units	Dimension
General heat capacity	C = heat capacity of substance	$q = CT$	$J K^{-1}$	$[M][L]^2[T]^{-2}[\Theta]^{-1}$
Linear thermal expansion	L = length of material (m) α = coefficient linear thermal expansion (dimensionless) ε = strain tensor (dimensionless)	$\partial L/\partial T = \alpha L$ $\varepsilon_{ij} = \alpha_{ij}\Delta T$	K^{-1}	$[\Theta]^{-1}$
Volumetric thermal expansion	β, γ V = volume of object (m^3) p = constant pressure of surroundings	$(\partial V/\partial T)_p = \gamma V$	K^{-1}	$[\Theta]^{-1}$
Thermal conductivity	κ, K, λ, **A** = surface cross section of material (m^2) P = thermal current/power through material (W) ∇T = temperature gradient in material (K m $^{-1}$)	$\lambda = -P/(\mathbf{A} \cdot \nabla T)$	$W m^{-1} K^{-1}$	$[M][L][T]^{-3}[\Theta]^{-1}$
Thermal conductance	U	$U = \lambda/\delta x$	$W m^{-2} K^{-1}$	$[M][T]^{-3}[\Theta]^{-1}$
Thermal resistance	R Δx = displacement of heat transfer (m)	$R = 1/U = \Delta x/\lambda$	$m^2 K W^{-1}$	$[M]^{-1}[L][T]^3[\Theta]$

Definitions (Electrical/magnetic properties of matter)

Quantity (Common Name/s)	(Common) Symbol/s	Defining Equation	SI Units	Dimension
Electrical resistance	R	$R = V/I$	$\Omega = V A^{-1} = J s C^{-2}$	$[M] [L]^2 [T]^{-3} [I]^{-2}$
Resistivity	ρ	$\rho = RA/l$	Ω m	$[M]^2 [L]^2 [T]^{-3} [I]^{-2}$
Resistivity temperature coefficient, linear temperature dependence	α	$\rho - \rho_0 = \rho_0\alpha(T - T_0)$	K^{-1}	$[\Theta]^{-1}$
Electrical conductance	G	$G = 1/R$	$S = \Omega^{-1}$	$[T]^3 [I]^2 [M]^{-1} [L]^{-2}$
Electrical conductivity	σ	$\sigma = 1/\rho$	$\Omega^{-1} m^{-1}$	$[I]^2 [T]^3 [M]^{-2} [L]^{-2}$
Magnetic reluctance	$R, R_m, \mathcal{R}$	$R_m = \mathcal{M}/\Phi_B$	$A Wb^{-1} = H^{-1}$	$[M]^{-1}[L]^{-2}[T]^2$
Magnetic permeance	$P, P_m, \Lambda, \mathcal{P}$	$\Lambda = 1/R_m$	$Wb A^{-1} = H$	$[M][L]^2[T]^{-2}$

Definitive Laws

There are several laws which describe the transport of matter, or properties of it, in an almost identical way. In every case, in words they read:

Flux (density) is proportional to a ***gradient***, the constant of proportionality is the characteristic of the material.

In general the constant must be replaced by a 2nd rank tensor, to account for directional dependences of the material.

Property/effect	Nomenclature	Equation
Fick's law of diffusion, defines diffusion coefficient D	D = mass diffusion coefficient ($m^2 s^{-1}$) J = diffusion flux of substance (mol $m^{-2} s^{-1}$) $\partial C/\partial x$ = (1d)concentration gradient of substance (mol dm^{-4})	$J_i = -D_{ij}\frac{\partial C}{\partial x_j}$
Darcy's law for fluid flow in porous media, defines permeability κ	κ = permeability of medium (m^2) μ = fluid viscosity (Pa s) q = discharge flux of substance (m s^{-1}) $\partial P/\partial x$ = (1d) pressure gradient of system (Pa m^{-1})	$q_j = -\frac{\kappa}{\mu}\frac{\partial P}{\partial x_j}$
Ohm's law of electric conduction, defines electric conductivity (and hence resistivity and resistance)	V = potential difference in material (V) I = electric current through material (A) R = resistance of material (Ω) $\partial V/\partial x$ = potential gradient (electric field) through material (V m^{-1}) J = electric current density through material (A m^{-2}) σ = electric conductivity of material ($\Omega^{-1} m^{-1}$) ρ = electrical resistivity of material (Ω m)	Simplist form is: $V = IR$ More general forms are: $\frac{\partial V}{\partial x_j} = \rho_{ji} J_i \rightleftharpoons J_i = \sigma_{ij}\frac{\partial V}{\partial x_j}$
Fourier's law of thermal conduction, defines thermal conductivity λ	λ = thermal conductivity of material (W $m^{-1} K^{-1}$) q = heat flux through material (W m^{-2}) $\partial T/\partial x$ = temperature gradient in material (K m^{-1})	$q_i = -\lambda_{ij}\frac{\partial T}{\partial x_j}$
Stefan–Boltzmann law of black-body radiation, defines emmisivity ε	I = radiant intensity (W m^{-2}) σ = Stefan–Boltzmann constant (W $m^{-2} K^{-4}$) T_{sys} = temperature of radiating system (K) T_{ext} = temperature of external surroundings (K) ε = emissivity (dimensionless)	For a single radiator: $I = \varepsilon\sigma T^4$ For a temperature difference: $I = \varepsilon\sigma(T_{ext}^4 - T_{sys}^4)$ $0 \le \varepsilon \le 1$ ε = 0 for perfect reflector ε = 1 for perfect absorber (true black body)

CONSTRAINT (MATHEMATICS)

In mathematics, a **constraint** is a condition of an optimization problem that the solution must satisfy. There are several types of constraints—primarily equality constraints, inequality constraints, and integer constraints. The set of candidate solutions that satisfy all constraints is called the feasible set.

Example

The following is a simple optimization problem:

$$\min f(\mathbf{x}) = x_1^2 + x_2^4$$

subject to

$$x_1 \geq 1$$

and

$$x_2 = 1,$$

Where X denotes the vector (x_1, x_2).

In this example, the first line defines the function to be minimized (called the objective function, loss function, or cost function). The second and third lines define two constraints, the first of which is an inequality constraint and the second of which is an equality constraint. These two constraints are hard constraints, meaning that it is required that they be satisfied; they define the feasible set of candidate solutions.

Without the constraints, the solution would be (0,0), where $f(x)$ has the lowest value. But this solution does not satisfy the constraints. The solution of the constrained optimization problem stated above is $x = (1,1)$, which is the point with the smallest value of $f(x)$ that satisfies the two constraints.

Terminology

- If an inequality constraint holds with equality at the optimal point, the constraint is said to be **binding**, as the point cannot be varied in the direction of the constraint even though doing so would improve the value of the objective function.
- If an inequality constraint holds as a strict inequality at the optimal point (that is, does not hold with equality), the constraint is said to be **non-binding**, as the point could be varied in the direction of the constraint, although it would not be optimal to do so. If a constraint is non-binding, the optimization problem would have the same solution even in the absence of that constraint.
- If a constraint is not satisfied at a given point, the point is said to be **infeasible**.

Hard and Soft Constraints

If the problem mandates that the constraints be satisfied, as in the above discussion, the constraints are sometimes referred to as hard constraints. However, in some problems, called flexible constraint satisfaction problems, it is preferred but not required that certain constraints be satisfied; such non-mandatory constraints are known as soft constraints. Soft constraints arise in, for example, preference-based planning. In a MAX-CSP problem, a number of constraints are allowed to be violated, and the quality of a solution is measured by the number of satisfied constraints.

Classifications

Mathematical models are usually composed of relationships and variables. Relationships can be described by operators, such as algebraic operators, functions, differential operators, etc. Variables are abstractions of system parameters of interest, that can be quantified. Several classification criteria can be used for mathematical models according to their structure:

- **Linear vs. nonlinear:** If all the operators in a mathematical model exhibit linearity, the resulting mathematical model is defined as linear. A model is considered to be nonlinear otherwise. The definition of linearity and nonlinearity is dependent on context, and linear models may have nonlinear expressions in them. For example, in a statistical linear model, it is assumed that a relationship is linear in the parameters, but it may be nonlinear in the predictor variables.

Similarly, a differential equation is said to be linear if it can be written with linear differential operators, but it can still have nonlinear expressions in it. In a mathematical programming model, if the objective functions and constraints are represented entirely by linear equations, then the model is regarded as a linear model. If one or more of the objective functions or constraints are represented with a nonlinear equation, then the model is known as a nonlinear model. Nonlinearity, even in fairly simple systems, is often associated with phenomena such as chaos and irreversibility. Although there are exceptions, nonlinear systems and models tend to be more difficult to study than linear ones. A common approach to nonlinear problems is linearization, but this can be problematic if one is trying to study aspects such as irreversibility, which are strongly tied to nonlinearity.

- **Static vs. dynamic:** A dynamic model accounts for time-dependent changes in the state of the system, while a static (or steady-state) model calculates the system in equilibrium, and thus is time-invariant. Dynamic models typically are represented by differential equations or difference equations.
- **Explicit vs. implicit:** If all of the input parameters of the overall model are known, and the output parameters can be calculated by a finite series of computations, the model is said to be explicit. But sometimes it is the output parameters which are known, and the corresponding inputs must be solved for by an iterative procedure, such as Newton's method (if the model is linear) or Broyden's method (if non-linear). In such a case the model is said to be implicit. For example, a jet engine's physical properties such as turbine and nozzle throat areas can be explicitly calculated given a design thermodynamic cycle (air and fuel flow rates, pressures, and temperatures) at a specific flight condition and power setting, but the engine's operating cycles at other flight conditions and power settings cannot be explicitly calculated from the constant physical properties.
- **Discrete vs. continuous:** A discrete model treats objects as discrete, such as the particles in a molecular model or the states in a statistical model; while a continuous model represents the objects in a continuous manner, such as the velocity field of fluid in pipe flows, temperatures and stresses in a solid, and electric field that applies continuously over the entire model due to a point charge.
- **Deterministic vs. probabilistic (stochastic):** A deterministic model is one in which every set of variable states is uniquely determined by parameters in the model and by sets of previous states of these variables; therefore, a deterministic model always performs the same way for a given set of initial conditions. Conversely, in a stochastic model—usually called a "statistical model"—randomness is present, and variable states are not described by unique values, but rather by probability distributions.
- **Deductive, inductive, or floating:** A deductive model is a logical structure based on a theory. An inductive model arises from empirical findings and generalization from them. The floating model rests on neither theory nor observation, but is merely the invocation of expected structure. Application of mathematics in social sciences outside of economics has been criticized for unfounded models.

Application of catastrophe theory in science has been characterized as a floating model.

Significance in the Natural Sciences

Mathematical models are of great importance in the natural sciences, particularly in physics. Physical theories are almost invariably expressed using mathematical models.

Throughout history, more and more accurate mathematical models have been developed. Newton's laws accurately describe many everyday phenomena, but at certain limits relativity theory and quantum mechanics must be used; even these do not apply to all situations and need further refinement. It is possible to obtain the less accurate models in appropriate limits, for example relativistic mechanics reduces to Newtonian mechanics at speeds much less than the speed of light. Quantum mechanics reduces to classical physics when the quantum numbers are high. For example, the de Broglie wavelength of a tennis ball is insignificantly small, so classical physics is a good approximation to use in this case.

It is common to use idealized models in physics to simplify things. Massless ropes, point particles, ideal gases and the particle in a box are among the many simplified models used in physics. The laws of physics are represented with simple equations such as Newton's laws, Maxwell's equations and the Schrödinger equation. These laws are such as a basis for making mathematical models of real situations. Many real situations are very complex and thus modeled approximate on a computer, a model that is computationally feasible to compute is made from the basic laws or from approximate models made from the basic laws. For example, molecules can be modeled by molecular orbital models that are approximate solutions to the Schrödinger equation. In engineering, physics models are often made by mathematical methods such as finite element analysis.

Different mathematical models use different geometries that are not necessarily accurate descriptions of the geometry of the universe. Euclidean geometry is much used in classical physics, while special relativity and general relativity are examples of theories that use geometries which are not Euclidean.

Some Applications

Since prehistorical times simple models such as maps and diagrams have been used.

Often when engineers analyze a system to be controlled or optimized, they use a mathematical model. In analysis, engineers can build a descriptive model of the system as a hypothesis of how the system could work, or try to estimate how an unforeseeable event could affect the system. Similarly, in control of a system, engineers can try out different control approaches in simulations.

A mathematical model usually describes a system by a set of variables and a set of equations that establish relationships between the variables. Variables may be of many types; real or integer numbers, boolean values or strings, for example. The variables represent some properties of the system, for example, measured system outputs often in the form of signals, timing data, counters, and event occurrence (yes/no). The actual model is the set of functions that describe the relations between the different variables.

Building Blocks

In business and engineering, mathematical models may be used to maximize a certain output. The system under consideration will require certain inputs. The system relating inputs to outputs depends on other variables too: decision variables, state variables, exogenous variables, and random variables.

Decision variables are sometimes known as independent variables. Exogenous variables are sometimes known as parameters or constants. The variables are not independent of each other as the state variables are dependent on the decision, input, random, and exogenous variables. Furthermore, the output variables are dependent on the state of the system (represented by the state variables).

Objectives and constraints of the system and its users can be represented as functions of the output variables or state variables. The objective functions will depend on the perspective of the model's user. Depending on the context, an objective function is also known as an index of performance, as it is some measure of interest to the user. Although there is no limit to the number of objective functions and constraints a model can have, using or optimizing the model becomes more involved (computationally) as the number increases.

For example, economists often apply linear algebra when using input-output models. Complicated mathematical models that have many variables may be consolidated by use of vectors where one symbol represents several variables.

A Priori Information

To analyse something with a typical "black box approach", only the behavior of the stimulus/response will be accounted for, to infer the (unknown) box. The usual representation of this black box system is a data flow diagram centered in the box.

Mathematical modeling problems are often classified into black box or white box models, according to how much a priori information on the system is available. A black-box model is a system of which there is no a priori information available. A white-box model (also called glass box or clear box) is a system where all necessary information is available. Practically all systems are somewhere between the black-box and white-box models, so this concept is useful only as an intuitive guide for deciding which approach to take.

Usually it is preferable to use as much a priori information as possible to make the model more accurate. Therefore, the white-box models are usually considered easier, because if you have used the information correctly, then the model will behave correctly. Often the a priori information comes in forms of knowing the type of functions relating different variables. For example, if we make a model of how a medicine works in a human system, we know that usually the amount of medicine in the blood is an exponentially decaying function. But we are still left with several unknown parameters; how rapidly does the medicine amount decay, and what is the initial amount of medicine in blood? This example is therefore not a completely white-box model. These parameters have to be estimated through some means before one can use the model.

In black-box models one tries to estimate both the functional form of relations between variables and the numerical parameters in those functions. Using a priori information we could end up, for example, with a set of functions that probably could describe the system adequately. If there is no a priori information we would try to use functions as general as possible to cover all different models. An often used approach for black-box models are neural networks which usually do not make assumptions about incoming data. Alternatively the NARMAX (Nonlinear AutoRegressive Moving Average model with eXogenous inputs) algorithms which were developed as part of nonlinear system identification can be used to select the model terms, determine the model structure, and estimate the unknown parameters in the presence of correlated and nonlinear noise. The advantage of NARMAX models compared to neural networks is that NARMAX produces models that can be written down and related to the underlying process, whereas neural networks produce an approximation that is opaque.

Subjective Information

Sometimes it is useful to incorporate subjective information into a mathematical model. This can be done based on intuition, experience, or expert opinion, or based on convenience of mathematical form. Bayesian statistics provides a theoretical framework for incorporating such subjectivity into a rigorous analysis: we specify a prior probability distribution (which can be subjective), and then update this distribution based on empirical data.

An example of when such approach would be necessary is a situation in which an experimenter bends a coin slightly and tosses it once, recording whether it comes up heads, and is then given the task of predicting the probability that the next flip comes up heads. After bending the coin, the true probability that the coin will come up heads is unknown; so the experimenter would need to make a decision (perhaps by looking at the shape of the coin) about what prior distribution to use. Incorporation of such subjective information might be important to get an accurate estimate of the probability.

Complexity

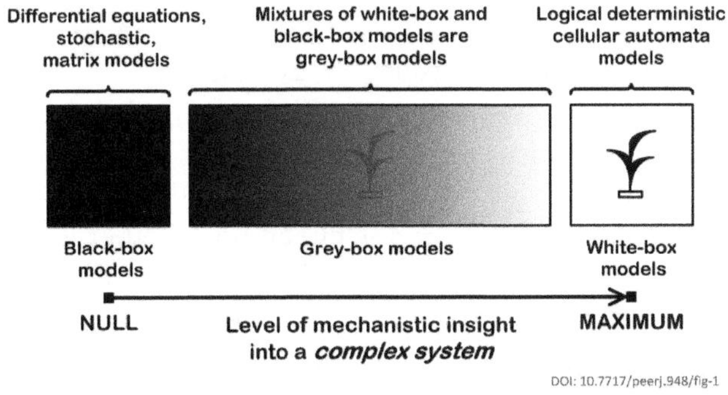

This is a schematic representation of three types of mathematical models of complex systems with the level of their mechanistic understanding.

In general, model complexity involves a trade-off between simplicity and accuracy of the model. Occam's razor is a principle particularly relevant to modeling, its essential idea being that among models with roughly equal predictive power, the simplest one is the most desirable. While added complexity usually improves the realism of a model, it can make the model difficult to understand and analyze, and can also pose computational problems, including numerical instability. Thomas Kuhn argues that as science progresses, explanations tend to become more complex before a paradigm shift offers radical simplification.

For example, when modeling the flight of an aircraft, we could embed each mechanical part of the aircraft into our model and would thus acquire an almost white-box model of the system. However, the computational cost of adding such a huge amount of detail would effectively inhibit the usage of such a model. Additionally, the uncertainty would increase due to an overly complex system, because each separate part induces some amount of variance into the model. It is therefore usually appropriate to make some approximations to reduce the model to a sensible size. Engineers often can accept some approximations in order to get a more robust and simple model. For example, Newton's classical mechanics is an approximated model of the real world. Still, Newton's model is quite sufficient for most ordinary-life situations, that is, as long as particle speeds are well below the speed of light, and we study macro-particles only.

Training

Any model which is not pure white-box contains some parameters that can be used to fit the model to the system it is intended to describe. If the modeling is done by a neural network or other machine learning, the optimization of parameters is called training, while the optimization of model hyperparameters is called tuning and often uses cross-validation. In more conventional modeling through explicitly given mathematical functions, parameters are often determined by curve fitting.

Model Evaluation

A crucial part of the modeling process is the evaluation of whether or not a given mathematical model describes a system accurately. This question can be difficult to answer as it involves several different types of evaluation.

Fit to Empirical Data

Usually the easiest part of model evaluation is checking whether a model fits experimental measurements or other empirical data. In models with parameters, a common approach to test this fit is to split the data into two disjoint subsets: training data and verification data. The training data are used to estimate the model parameters. An accurate model will closely match the verification data even though these data were not used to set the model's parameters. This practice is referred to as cross-validation in statistics.

Defining a metric to measure distances between observed and predicted data is a useful tool of assessing model fit. In statistics, decision theory, and some economic models, a loss function plays a similar role.

While it is rather straightforward to test the appropriateness of parameters, it can be more difficult to test the validity of the general mathematical form of a model. In general, more mathematical tools have been developed to test the fit of statistical models than models involving differential equations. Tools from non-parametric statistics can sometimes be used to evaluate how well the data fit a known distribution or to come up with a general model that makes only minimal assumptions about the model's mathematical form.

Scope of the Model

Assessing the scope of a model, that is, determining what situations the model is applicable to, can be less straightforward. If the model was constructed based on a set of data, one must determine for which systems or situations the known data is a "typical" set of data.

The question of whether the model describes well the properties of the system between data points is called interpolation, and the same question for events or data points outside the observed data is called extrapolation.

As an example of the typical limitations of the scope of a model, in evaluating Newtonian classical mechanics, we can note that Newton made his measurements without advanced equipment, so he could not measure properties of particles travelling at speeds close to the speed of light. Likewise, he did not measure the movements of molecules and other small particles, but macro particles only. It is then not surprising that his model does not extrapolate well into these domains, even though his model is quite sufficient for ordinary life physics.

Philosophical Considerations

Many types of modeling implicitly involve claims about causality. This is usually (but not always) true of models involving differential equations. As the purpose of modeling is to increase our understanding of the world, the validity of a model rests not only on its fit to empirical observations, but also on its ability to extrapolate to situations or data beyond those originally described in the model. One can think of this as the differentiation between qualitative and quantitative predictions. One can also argue that a model is worthless unless it provides some insight which goes beyond what is already known from direct investigation of the phenomenon being studied.

An example of such criticism is the argument that the mathematical models of Optimal foraging theory do not offer insight that goes beyond the common-sense conclusions of evolution and other basic principles of ecology.

Examples

- One of the popular examples in computer science is the mathematical models of various machines, an example is the deterministic finite automaton which is defined as an abstract mathematical concept, but due to the deterministic nature of a DFA, it is implementable in hardware and software for solving various specific problems. For example, the following is a DFA M with a binary alphabet, which requires that the input contains an even number of 0s.

REFERENCES

1. Andreski, Stanislav (1972). Social Sciences as Sorcery. St. Martin's Press. ISBN 0-14-021816-5.
2. Truesdell, Clifford (1984). An Idiot's Fugitive Essays on Science. Springer. pp. 121–7. ISBN 3-540-90703-3.
3. Billings S.A. (2013), Nonlinear System Identification: NARMAX Methods in the Time, Frequency, and Spatio-Temporal Domains, Wiley.
4. Pyke, G. H. (1984). "Optimal Foraging Theory: A Critical Review". Annual Review of Ecology and Systematics. 15: 523–575. doi:10.1146/annurev.es.15.110184.002515.
5. landinfo.com, definition of map projection
6. Gallistel (1990). The Organization of Learning. Cambridge: The MIT Press. ISBN 0-262-07113-4.
7. Whishaw, I. Q.; Hines, D. J.; Wallace, D. G. (2001). "Dead reckoning (path integration) requires the hippocampal formation: Evidence from spontaneous exploration and spatial learning tasks in light (allothetic) and dark (idiothetic) tests". Behavioural Brain Research. 127 (1–2): 49–69. doi:10.1016/S0166-4328(01)00359-X. PMID 11718884.

CHAPTER 6

Potential theory

Chapter Contents

In mathematics and mathematical physics, potential theory is the study of harmonic functions. The term "potential theory" was coined in 19th-century physics, when it was realized that the fundamental forces of nature could be modeled using potentials which satisfy Laplace's equation.

INTRODUCTION

In mathematics and mathematical physics, potential theory is the study of harmonic functions.

The term "potential theory" was coined in 19th-century physics, when it was realized that the fundamental forces of nature could be modeled using potentials which satisfy Laplace›s equation. Although more accurate theories - for example classical electrostatics and Newtonian gravity- were developed later, the name "potential theory" remained.

There is considerable overlap between potential theory and the theory of the Laplace equation. To the extent that it is possible to draw a distinction between these two fields, the difference is more one of emphasis than subject matter and rests on the following distinction: potential theory focuses on the properties of the functions as opposed to the properties of the equation. For example, a result about the singularities of harmonic functions would be said to belong to potential theory whilst a result on how the solution depends on the boundary data would be said to belong to the theory of the Laplace equation. Of course, this is not a hard and fast distinction, and in practice there is considerable overlap between the two fields, with methods and results from one being used in the other.

Modern potential theory is also intimately connected with probability and the theory of Markov chains. In the continuous case, this is closely related to analytic theory. In the finite state space case, this connection can be introduced by introducing an electrical network on the state space, with resistance between points inversely proportional to transition probabilities and densities proportional to potentials. Even in the finite case, the analogue I-K of the Laplacian in potential theory has its own maximum principle, uniqueness principle, balance principle, and others.

SYMMETRY

A useful starting point and organizing principle in the study of harmonic functions is a consideration of the symmetries of the Laplace equation. Although it is not a symmetry in the usual sense of the term, we can start with the observation that the Laplace equation is linear. This means that the fundamental object of study in potential theory is a linear space of functions. This observation will prove especially important when we consider function space approaches to the subject in a later section.

As for symmetry in the usual sense of the term, we may start with the theorem that the symmetries of the n dimensional Laplace equation are exactly the conformal symmetries of the n-dimensional Euclidean space. This fact has several implications. First of all, one can consider harmonic functions which transform under irreducible representations of the conformal group or of its subgroups (such as the group of rotations or translations). Proceeding in this fashion, one systematically obtains the solutions of the Laplace equation which arise from separation of variables such as spherical harmonic solutions and Fourier series. By taking linear superpositions of these solutions, one can produce large classes of harmonic functions which can be shown to be dense in the space of all harmonic functions under suitable topologies.

Second, one can use conformal symmetry to understand such classical tricks and techniques for generating harmonic functions as the Kelvin transform and the method of images.

Third, one can use conformal transforms to map harmonic functions in one domain to harmonic functions in another domain. The most common instance of such a construction is to relate harmonic functions on a disk to harmonic functions on a half-plane.

Fourth, one can use conformal symmetry to extend harmonic functions to harmonic functions on conformally flat Riemannian manifolds. Perhaps the simplest such extension is to consider a harmonic function defined on the whole of $\mathbf{R}^n$ (with the possible exception of a discrete set of singular points) as a harmonic function on the n-dimensional sphere. More complicated situations can also happen. For instance, one can obtain a higher-dimensional analog of Riemann surface theory by expressing a multi-valued harmonic function as a single-valued function on a branched cover of $\mathbf{R}^n$ or one can regard harmonic functions which are invariant under a discrete subgroup of the conformal group as functions on a multiply connected manifold or orbifold.

Two Dimensions

From the fact that the group of conformal transforms is infinite-dimensional in two dimensions and finite-dimensional for more than two dimensions, one can surmise that potential theory in two dimensions is different from potential theory in other dimensions. This is correct and, in fact, when one realizes that any two-dimensional harmonic function is the real part of a complex analytic function, one sees that the subject of two-dimensional potential theory is substantially the same as that of complex analysis. For this reason, when speaking of potential theory, one focuses attention on theorems which hold in three or more dimensions. In this connection, a surprising fact is that many results and concepts originally discovered in complex analysis (such as Schwarz's theorem, Morera's theorem, the Weierstrass-Casorati theorem, Laurent series, and the classification of singularities as removable, poles and essential singularities) generalize to results on harmonic functions in any dimension. By considering which theorems of complex analysis are special cases of theorems of potential theory in any dimension, one can obtain a feel for exactly what is special about complex analysis in two dimensions and what is simply the two-dimensional instance of more general results.

Local Behavior

An important topic in potential theory is the study of the local behavior of harmonic functions. Perhaps the most fundamental theorem about local behavior is the regularity theorem for Laplace's equation, which states that harmonic functions are analytic. There are results which describe the local structure of level sets of harmonic functions. There is Bôcher's theorem, which characterizes the behavior of isolated singularities of positive harmonic functions. As alluded to in the last section, one can classify the isolated singularities of harmonic functions as removable singularities, poles, and essential singularities.

Inequalities

A fruitful approach to the study of harmonic functions is the consideration of inequalities they satisfy. Perhaps the most basic such inequality, from which most other inequalities may be derived, is the maximum principle. Another important result is Liouville's theorem, which states the only bounded harmonic functions defined on the whole of $\mathbf{R}^n$ are, in fact, constant functions. In addition to these basic inequalities, one has Harnack›s inequality, which states that positive harmonic functions on bounded domains are roughly constant.

One important use of these inequalities is to prove convergence of families of harmonic functions or sub-harmonic functions, see Harnack›s theorem. These convergence theorems are used to prove the existence of harmonic functions with particular properties.

Spaces of Harmonic Functions

Since the Laplace equation is linear, the set of harmonic functions defined on a given domain is, in fact, a vector space. By defining suitable norms and/or inner products, one can exhibit sets of harmonic functions which form Hilbert or Banach spaces. In this fashion, one obtains such spaces as the Hardy space, Bloch space, and Bergman space.

HARMONIC FUNCTION

In mathematics, mathematical physics and the theory of stochastic processes, a **harmonic function** is a twice continuously differentiable function $f: U \to \mathbf{R}$ (where U is an open subset of $\mathbf{R}^n$) which satisfies Laplace's equation, i.e.

$$\frac{\partial^2 f}{\partial x_1^2} + \frac{\partial^2 f}{\partial x_2^2} + \cdots + \frac{\partial^2 f}{\partial x_n^2} = 0$$

everywhere on U. This is usually written as

$$\nabla^2 f = 0$$

or

$$\Delta f = 0$$

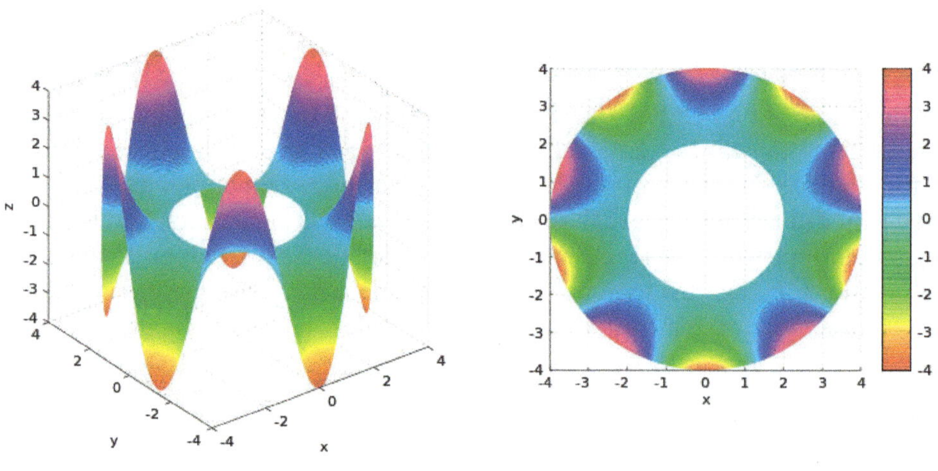

A harmonic function defined on an annulus.

Etymology of the Term "Harmonic"

The descriptor "harmonic" in the name harmonic function originates from a point on a taut string which is undergoing harmonic motion. This solution to the differential equation for this type of motion can be written in terms of sines and cosines, functions which are thus referred to as *harmonics*. Fourier analysis involves expanding periodic functions on the unit circle in terms of a series of these harmonics. Considering higher dimensional analogues of the harmonics on the unit n-sphere, one arrives at the spherical harmonics. These functions satisfy Laplace's equation and over time "harmonic" was used to refer to all functions satisfying Laplace's equation.

Examples

Examples of harmonic functions of two variables are:

- The real or imaginary part of any holomorphic function
- The function $f(x, y) = e^x \sin y$; this is a special case of the example above, as $f(x, y) = \operatorname{Im}(e^{x+iy})$, and e^{x+iy} is a holomorphic function.
- The function $f(x_1, x_2) = \ln(x_1^2 + x_2^2)$ defined on $\mathbb{R}^2 \setminus \{0\}$ (e.g. the electric potential due to a line charge, and the gravity potential due to a long cylindrical mass)

Examples of harmonic functions of three variables are given in the table below with $r^2 = x^2 + y^2 + z^2$:

Function	Singularity
$\frac{1}{r}$	Unit point charge at origin
$\frac{x}{r^3}$	x-directed dipole at origin
$-\ln(r^2 - z^2)$	Line of unit charge density on entire z-axis
$-\ln(r + z)$	Line of unit charge density on negative z-axis
$\frac{x}{r^2 - z^2}$	Line of x-directed dipoles on entire z axis
$\frac{x}{r(r + z)}$	Line of x-directed dipoles on negative z axis

Harmonic functions that arise in physics are determined by their singularities and boundary conditions (such as Dirichlet boundary conditions or Neumann boundary conditions). On regions without boundaries, adding the real or imaginary part of any entire function will produce a harmonic function with the same singularity, so in this case the harmonic function is not determined by its singularities; however, we can make the solution unique in physical situations by requiring that the solution goes to 0 as you go to infinity. In this case, uniqueness follows by Liouville's theorem.

The singular points of the harmonic functions above are expressed as "charges" and "charge densities" using the terminology of electrostatics, and so the corresponding harmonic function will be proportional to the electrostatic potential due to these charge distributions. Each function above will yield another harmonic function when multiplied by a constant, rotated, and/or has a constant added. The inversion of each function will yield another harmonic function which has singularities which are the images of the original singularities in a spherical "mirror". Also, the sum of any two harmonic functions will yield another harmonic function.

Finally, examples of harmonic functions of n variables are:

- The constant, linear and affine functions on all of $\mathbf{R}^n$ (for example, the electric potential between the plates of a capacitor, and the gravity potential of a slab)
- The function $f(x_1,\dots,x_n) = ({x_1}^2 + \cdots + {x_n}^2)^{1-n/2}$ on $\mathbb{R}^n \setminus \{0\}$ for $n > 2$.

Remarks

The set of harmonic functions on a given open set U can be seen as the kernel of the Laplace operator Δ and is therefore a vector space over $\mathbf{R}$: sums, differences and scalar multiples of harmonic functions are again harmonic.

If f is a harmonic function on U, then all partial derivatives of f are also harmonic functions on U. The Laplace operator Δ and the partial derivative operator will commute on this class of functions.

In several ways, the harmonic functions are real analogues to holomorphic functions. All harmonic functions are analytic, i.e. they can be locally expressed as power series. This is a general fact about elliptic operators, of which the Laplacian is a major example.

The uniform limit of a convergent sequence of harmonic functions is still harmonic. This is true because every continuous function satisfying the mean value property is harmonic. Consider the sequence on $(-\infty, 0) \times \mathbf{R}$ defined by $f_n(x,y) = \frac{1}{n}\exp(nx)\cos(ny)$.

This sequence is harmonic and converges uniformly to the zero function; however note that the partial derivatives are not uniformly convergent to the zero function (the derivative of the zero function). This example shows the importance of relying on the mean value property and continuity to argue that the limit is harmonic.

Connections with Complex Function Theory

The real and imaginary part of any holomorphic function yield harmonic functions on $\mathbf{R}^2$ (these are said to be a pair of harmonic conjugate functions). Conversely, any harmonic function u on an open subset Ω of $\mathbf{R}^2$ is *locally* the real part of a holomorphic function. This is immediately seen observing that, writing $z = x + iy$, the complex function $g(z) := u_x - \mathrm{i}\, u_y$ is holomorphic in Ω because it satisfies the Cauchy–Riemann equations. Therefore, g has locally a primitive f, and u is the real part of f up to a constant, as u_x is the real part of $f'=g$.

Although the above correspondence with holomorphic functions only holds for functions of two real variables, harmonic functions in n variables still enjoy a number of properties typical of holomorphic functions. They are (real) analytic; they have a maximum principle and a mean-value principle; a theorem of removal of singularities as well as a Liouville theorem holds for them in analogy to the corresponding theorems in complex functions theory.

Properties of Harmonic Functions

Some important properties of harmonic functions can be deduced from Laplace's equation.

Regularity Theorem for Harmonic Functions

Harmonic functions are infinitely differentiable. In fact, harmonic functions are real analytic.

Maximum Principle

Harmonic functions satisfy the following *maximum principle*: if K is any compact subset of U, then f, restricted to K, attains its maximum and minimum on the boundary of K. If U is connected, this means that f cannot have local maxima or minima, other than the exceptional case where f is constant. Similar properties can be shown for subharmonic functions.

The Mean Value Property

If $B(x, r)$ is a ball with center x and radius r which is completely contained in the open set $\Omega \subset \mathbf{R}^n$, then the value $u(x)$ of a harmonic function u: $\Omega \to \mathbf{R}$ at the center of the ball is given by the average value of u on the surface of the ball; this average value is also equal to the average value of u in the interior of the ball. In other words,

$$u(x) = \frac{1}{n\omega_n r^{n-1}} \int_{\partial B(x,r)} u \, d\sigma = \frac{1}{\omega_n r^n} \int_{B(x,r)} u \, dV$$

where ωn is the area of the unit sphere in n dimensions and σ is the n-1 dimensional surface measure .

Conversely, all locally integrable functions satisfying the (volume) mean-value property are both infinitely differentiable and harmonic.

In terms of convolutions, if

$$\chi_r := \frac{1}{|B(0,r)|}\chi_{B(0,r)} = \frac{n}{\omega_n r^n}\chi_{B(0,r)}$$

denotes the characteristic function of the ball with radius r about the origin, normalized so that ∫ R n χ r d x = 1 , the function u is harmonic on Ω if and only if

$$u(x) = u * \chi_r(x)$$

as soon as $B(x, r) \subset \Omega$.

Sketch of the proof. The proof of the mean-value property of the harmonic functions and its converse follows immediately observing that the non-homogeneous equation, for any $0 < s < r$

$$\Delta w = \chi_r - \chi_s$$

admits an easy explicit solution wr,s of class C1,1 with compact support in B(0, r). Thus, if u is harmonic in Ω

$$0 = \Delta u * w_{r,s} = u * \Delta w_{r,s} = u * \chi_r - u * \chi_s$$

holds in the set Ω_r of all points x in Ω with $\operatorname{dist}(x, \partial\Omega) > r$.

This statement of the mean value property can be generalized as follows: If h is any spherically symmetric function supported in $B(x,r)$ such that $\int h = 1$, then $u(x) = h * u(x)$. In other words, we can take the weighted average of u about a point and recover $u(x)$. In particular, by taking h to be a C^∞function, we can recover the value of u at any point even if we only know how u acts as a distribution. See Weyl's lemma.

Harnack›s Inequality

Let u be a non-negative harmonic function in a bounded domain Ω. Then for every connected set

$$V \subset \overline{V} \subset \Omega,$$

Harnack's inequality

$$\sup_V u \leq C \inf_V u$$

Removal of Singularities

The following principle of removal of singularities holds for harmonic functions. If f is a harmonic function defined on a dotted open subset $\Omega \setminus \{x\}$of Rn which is less singular at x_0 than the fundamental solution, that is

$$f(x) = o\left(|x - x_0|^{2-n}\right), \qquad \text{as } x \to x_0,$$

then f extends to a harmonic function on Ω (compare Riemann›s theorem for functions of a complex variable).

Liouville's Theorem

If f is a harmonic function defined on all of $\mathbf{R}^n$ which is bounded above or bounded below, then f is constant (compare Liouville›s theorem for functions of a complex variable).

Edward Nelson gave a particularly short proof of this theorem, using the mean value property mentioned above:

Given two points, choose two balls with the given points as centers and of equal radius. If the radius is large enough, the two balls will coincide except for an arbitrarily small proportion of their volume. Since f is bounded, the averages of it over the two balls are arbitrarily close, and so f assumes the same value at any two points.

Generalizations

Weakly Harmonic Function

A function (or, more generally, a distribution) is weakly harmonic if it satisfies Laplace's equation

$\Delta f = 0,$

in a weak sense (or, equivalently, in the sense of distributions). A weakly harmonic function coincides almost everywhere with a strongly harmonic function, and is in particular smooth. A weakly harmonic distribution is precisely the distribution associated to a strongly harmonic function, and so also is smooth. This is Weyl›s lemma.

There are other weak formulations of Laplace›s equation that are often useful. One of which is Dirichlet›s principle, representing harmonic functions in the Sobolev space $H^1(\Omega)$ as the minimizers of the Dirichlet energy integral

$$J(u) := \int_\Omega |\nabla u|^2 \, dx$$

with respect to local variations, that is, all functions $u \in H^1(\Omega)$ such that $J(u) \leq J(u + v)$ holds for all $v \in C_c^\infty(\Omega)$, or equivalently, for all $v \in H_0^1(\Omega)$.

Harmonic Functions on Manifolds

Harmonic functions can be defined on an arbitrary Riemannian manifold, using the Laplace–Beltrami operator Δ. In this context, a function is called *harmonic* if

$\Delta f = 0,$

Many of the properties of harmonic functions on domains in Euclidean space carry over to this more general setting, including the mean value theorem (over geodesic balls), the maximum principle, and the Harnack inequality. With the exception of the mean value theorem, these are easy consequences of the corresponding results for general linear elliptic partial differential equations of the second order.

Subharmonic Functions

A C^2 function that satisfies $\Delta f \geq 0$ is called subharmonic. This condition guarantees that the maximum principle will hold, although other properties of harmonic functions may fail. More generally, a function is subharmonic if and only if, in the interior of any ball in its domain, its graph lies below that of the harmonic function interpolating its boundary values on the ball.

Harmonic Forms

One generalization of the study of harmonic functions is the study of harmonic forms on Riemannian manifolds, and it is related to the study of cohomology. Also, it is possible to define harmonic vector-valued functions, or harmonic maps of two Riemannian manifolds, which are critical points of a generalized Dirichlet energy functional (this includes harmonic functions as a special case, a result known as Dirichlet principle). This kind of harmonic maps appear in the theory of minimal surfaces. For example, a curve, that is, a map from an interval in **R** to a Riemannian manifold, is a harmonic map if and only if it is a geodesic.

Harmonic Maps Between Manifolds

If M and N are two Riemannian manifolds, then a harmonic map $u : M \to N$ is defined to be a critical point of the Dirichlet energy

$$D[u] = \frac{1}{2} \int_M \|du\|^2 \, d\mathrm{Vol}$$

in which $du : TM \to TN$ is the differential of u, and the norm is that induced by the metric on M and that on N on the tensor product bundle $T^*M \otimes u^{-1} TN$.

Important special cases of harmonic maps between manifolds include minimal surfaces, which are precisely the harmonic immersions of a surface into three-dimensional Euclidean space. More generally, minimal submanifolds are harmonic immersions of one manifold in another. Harmonic coordinates are a harmonic diffeomorphism from a manifold to an open subset of a Euclidean space of the same dimension.

REFERENCES

1. Garabedian, P. R.; Schiffer, M. (1950). "On existence theorems of potential theory and conformal mapping". Annals of Mathematics. 52 (1): 164–187. doi:10.2307/1969517. JSTOR 1969517. Axler, Sheldon; Bourdon, Paul; Ramey, Wade (2001). Harmonic Function Theory. New York: Springer. p. 25. ISBN 0-387-95218-7.
2. Edward Nelson, A proof of Liouville's theorem. Proceedings of the AMS, 1961. pdf at JSTOR

CHAPTER 7

Probability

Chapter Contents

Probability is a branch of mathematics that deals with calculating the likelihood of a given event's occurrence, which is expressed as a number between 1 and 0. An event with a probability of 1 can be considered a certainty: for example, the probability of a coin toss resulting in either "heads" or "tails" is 1, because there are no other options, assuming the coin lands flat. An event with a probability of .5 can be considered to have equal odds of occurring or not occurring: for example, the probability of a coin toss resulting in "heads" is .5, because the toss is equally as likely to result in "tails." An event with a probability of 0 can be considered an impossibility: for example, the probability that the coin will land (flat) without either side facing up is 0, because either "heads" or "tails" must be facing up. A little paradoxical, probability theory applies precise calculations to quantify uncertain measures of random events.

INTRODUCTION

Probability is the measure of the likelihood that an event will occur. Probability is quantified as a number between 0 and 1 (where 0 indicates impossibility and 1 indicates certainty). The higher the probability of an event, the more certain that the event will occur.

A simple example is the tossing of a fair (unbiased) coin. Since the coin is unbiased, the two outcomes ("head" and "tail") are both equally probable; the probability of "head" equals the probability of "tail." Since no other outcomes are possible, the probability is 1/2 (or 50%), of either "head" or "tail". In other words, the probability of "head" is 1 out of 2 outcomes and the probability of "tail" is also 1 out of 2 outcomes, expressed as 0.5 when converted to decimal, with the above-mentioned quantification system. This type of probability is also called a priori probability.

These concepts have been given an axiomatic mathematical formalization in probability theory, which is used widely in such areas of study as mathematics, statistics, finance, gambling, science (in particular physics), artificial intelligence/machine learning, computer science, game theory, and philosophy to, for example, draw inferences about the expected frequency of events. Probability theory is also used to describe the underlying mechanics and regularities of complex systems.

INTERPRETATIONS

When dealing with experiments that are random and well-defined in a purely theoretical setting (like tossing a fair coin), probabilities can be numerically described by the number of desired outcomes divided by the total number of all outcomes. For example, tossing a fair coin twice will yield "head-head", "head-tail", "tail-head", and "tail-tail" outcomes. The probability of getting an outcome of "head-head" is 1 out of 4 outcomes or 1/4 or 0.25 (or 25%). When it comes to practical application however, there are two major competing categories of probability interpretations, whose adherents possess different views about the fundamental nature of probability:

- Objectivists assign numbers to describe some objective or physical state of affairs. The most popular version of objective probability is frequentist probability, which claims that the probability of a random event denotes the *relative frequency of occurrence* of an experiment's outcome, when repeating the experiment. This interpretation considers probability to be the relative frequency "in the long run" of outcomes. A modification of this is propensity probability, which interprets probability as the tendency of some experiment to yield a certain outcome, even if it is performed only once.
- Subjectivists assign numbers per subjective probability, i.e., as a degree of belief. The degree of belief has been interpreted as, "the price at which you would buy or sell a bet that pays 1 unit of utility if E, 0 if not E." The most popular version of subjective probability is Bayesian probability, which includes expert knowledge as well as experimental data to produce probabilities. The expert knowledge is represented by some (subjective) prior probability distribution. These data are incorporated in a likelihood function. The product of the prior and the likelihood, normalized, results in a posterior probability distribution that incorporates all the information known to date. Starting from arbitrary, subjective probabilities for a group of agents, some Bayesians claim that all agents will eventually have sufficiently similar assessments of probabilities, given enough evidence.

PROBABILITY INTERPRETATIONS

The word **probability** has been used in a variety of ways since it was first applied to the mathematical study of games of chance. Does probability measure the real, physical tendency of something to occur or is it a measure of how strongly one believes it will occur, or does it draw on both these elements? In answering such questions, mathematicians interpret the probability values of probability theory.

There are two broad categories of **probability interpretations** which can be called "physical" and "evidential" probabilities. Physical probabilities, which are also called objective or frequency probabilities, are associated with random physical systems such as roulette wheels, rolling dice and radioactive atoms. In such systems, a given type of event (such as a die yielding a six) tends to occur at a persistent rate, or "relative frequency", in a long run of trials. Physical probabilities either explain, or are invoked to explain, these stable frequencies. The two main kinds of theory of physical probability are frequentist accounts (such as those of Venn, Reichenbach and von Mises) and propensity accounts (such as those of Popper, Miller, Giere and Fetzer).

Evidential probability, also called Bayesian probability, can be assigned to any statement whatsoever, even when no random process is involved, as a way to represent its subjective plausibility, or the degree to which the statement is supported by the available evidence. On most accounts, evidential probabilities are considered to be degrees of belief, defined in terms of dispositions to gamble at certain odds. The four main evidential interpretations are the classical (e.g. Laplace's) interpretation, the subjective interpretation (de Finetti and Savage), the epistemic or inductive interpretation (Ramsey, Cox) and the logical interpretation (Keynes and Carnap). There are also evidential interpretations of probability covering groups, which are often labelled as ‹intersubjective' (proposed by Gillies and Rowbottom).

Some interpretations of probability are associated with approaches to statistical inference, including theories of estimation and hypothesis testing. The physical interpretation, for example, is taken by followers of "frequentist" statistical methods, such as Ronald Fisher, Jerzy Neyman and Egon Pearson. Statisticians of the opposing Bayesian school typically accept the existence and importance of physical probabilities, but also consider the calculation of evidential probabilities to be both valid and necessary in statistics. This article, however, focuses on the interpretations of probability rather than theories of statistical inference.

The terminology of this topic is rather confusing, in part because probabilities are studied within a variety of academic fields. The word "frequentist" is especially tricky. To philosophers it refers to a particular theory of physical probability, one that has more or less been abandoned. To scientists, on the other hand, "frequentist probability" is just another name for physical (or objective) probability. Those who promote Bayesian inference view "frequentist statistics" as an approach to statistical inference that recognises only physical probabilities. Also the word "objective", as applied to probability, sometimes means exactly what "physical" means here, but is also used of evidential probabilities that are fixed by rational constraints, such as logical and epistemic probabilities.

It is unanimously agreed that statistics depends somehow on probability. But, as to what probability is and how it is connected with statistics, there has seldom been such complete disagreement and breakdown of communication since the Tower of Babel. Doubtless, much of the disagreement is merely terminological and would disappear under sufficiently sharp analysis.

—(Savage, 1954, p 2)

Philosophy

The **philosophy of probability** presents problems chiefly in matters of epistemology and the uneasy interface between mathematical concepts and ordinary language as it is used by non-mathematicians. Probability theory is an established field of study in mathematics. It has its origins in correspondence discussing the mathematics of games of chance between Blaise Pascal and Pierre de Fermat in the seventeenth century, and was formalized and rendered axiomatic as a distinct branch of mathematics by Andrey Kolmogorov in the twentieth century. In axiomatic form, mathematical statements about probability theory carry the same sort of epistemological confidence within the philosophy of mathematics as are shared by other mathematical statements.

The mathematical analysis originated in observations of the behaviour of game equipment such as playing cards and dice, which are designed specifically to introduce random and equalized elements; in mathematical terms, they are subjects of indifference. This is not the only way probabilistic statements are used in ordinary human language: when people say that "*it will probably rain*", they typically do not mean that the outcome of rain versus not-rain is a random factor that the odds currently favor; instead, such statements are perhaps better understood as qualifying their expectation of rain with a degree of confidence. Likewise, when it is written that "the most probable explanation" of the name of Ludlow, Massachusetts "is that it was named after Roger Ludlow", what is meant here is not that Roger Ludlow is favored by a random factor, but rather that this is the most plausible explanation of the evidence, which admits other, less likely explanations.

Thomas Bayes attempted to provide a logic that could handle varying degrees of confidence; as such, Bayesian probability is an attempt to recast the representation of probabilistic statements as an expression of the degree of confidence by which the beliefs they express are held.

Though probability initially had somewhat mundane motivations, its modern influence and use is widespread ranging from evidence-based medicine, through Six sigma, all the way to the Probabilistically checkable proof and the String theory landscape.

A summary of some interpretations of probability

Classical	Frequentist	Subjective	Propensity	
Main hypothesis	Principle of indifference	Frequency of occurrence	Degree of belief	Degree of causal connection
Conceptual basis	Hypothetical symmetry	Past data and reference class	Knowledge and intuition	Present state of system

Conceptual approach	Conjectural	Empirical	Subjective	Metaphysical
Single case possible	Yes	No	Yes	Yes
Precise	Yes	No	No	Yes
Problems	Ambiguity in principle of indifference	Circular definition	Reference class problem	Disputed concept

Classical Definition

The first attempt at mathematical rigour in the field of probability, championed by Pierre-Simon Laplace, is now known as the **classical definition**. Developed from studies of games of chance (such as rolling dice) it states that probability is shared equally between all the possible outcomes, provided these outcomes can be deemed equally likely. (3.1)

The theory of chance consists in reducing all the events of the same kind to a certain number of cases equally possible, that is to say, to such as we may be equally undecided about in regard to their existence, and in determining the number of cases favorable to the event whose probability is sought. The ratio of this number to that of all the cases possible is the measure of this probability, which is thus simply a fraction whose numerator is the number of favorable cases and whose denominator is the number of all the cases possible.

—Pierre-Simon Laplace, A Philosophical Essay on Probabilities

The classical definition of probability works well for situations with only a finite number of equally-likely outcomes.

This can be represented mathematically as follows: If a random experiment can result in N mutually exclusive and equally likely outcomes and if N_A of these outcomes result in the occurrence of the event A, the **probability of A** is defined by

$$P(A) = \frac{N_A}{N}.$$

There are two clear limitations to the classical definition. Firstly, it is applicable only to situations in which there is only a 'finite' number of possible outcomes. But some important random experiments, such as tossing a coin until it rises heads, give rise to an infinite set of outcomes. And secondly, you need to determine in advance that all the possible outcomes are equally likely without relying on the notion of probability to avoid circularity—for instance, by symmetry considerations.

Frequentism

For frequentists, the probability of the ball landing in any pocket can be determined only by repeated trials in which the observed result converges to the underlying probability *in the long run*.

Frequentists posit that the probability of an event is its relative frequency over time, (3.4) i.e., its relative frequency of occurrence after repeating a process a large number of times under similar conditions. This is also known as aleatory probability. The events are assumed to be governed by some random physical phenomena, which are either phenomena that are predictable, in principle, with sufficient information (see Determinism); or phenomena which are essentially unpredictable. Examples of the first kind include tossing dice or spinning a roulette wheel; an example of the second kind is radioactive decay. In the case of tossing a fair coin, frequentists say that the probability of getting a heads is 1/2, not because there are two equally likely outcomes but because repeated series of large numbers of trials demonstrate that the empirical frequency converges to the limit 1/2 as the number of trials goes to infinity.

If we denote by n_a the number of occurrences of an event A in n trials, then if $\lim_{n\to+\infty} \frac{n_a}{n} = p$ we say that $P(A) = p$.

The frequentist view has its own problems. It is of course impossible to actually perform an infinity of repetitions of a random experiment to determine the probability of an event. But if only a finite number of repetitions of the process are performed, different relative frequencies will appear in different series of trials. If these relative frequencies are to define the probability, the probability will be slightly different every time it is measured. But the real probability should be the same every time. If we acknowledge the fact that we only can measure a probability with some error of measurement attached, we still get into problems as the error of measurement can only be expressed as a probability, the very concept we are trying to define. This renders even the frequency definition circular; see for example "What is the Chance of an Earthquake?"

Logical, Epistemic, and Inductive Probability

It is widely recognized that the term "probability" is sometimes used in contexts where it has nothing to do with physical randomness. Consider, for example, the claim that the extinction of the dinosaurs was probably caused by a large meteorite hitting the earth. Statements such as "Hypothesis H is probably true" have been interpreted to mean that the (presently available) empirical evidence (E, say) supports H to a high degree. This degree of support of H by E has been called the logical probability of H given E, or the epistemic probability of H given E, or the inductive probability of H given E.

The differences between these interpretations are rather small, and may seem inconsequential. One of the main points of disagreement lies in the relation between probability and belief. Logical probabilities are conceived (for example in Keynes' Treatise on Probability) to be objective, logical relations between propositions (or sentences), and hence not to depend in any way upon belief. They are degrees of (partial) entailment, or degrees of logical consequence, not degrees of belief. (They do, nevertheless, dictate proper degrees of belief, as is discussed below.) Frank P. Ramsey, on the other hand, was skeptical about the existence of such objective logical relations and argued that (evidential) probability is "the logic of partial belief". (p 157) In other words, Ramsey held that epistemic probabilities simply *are* degrees of rational belief, rather than being logical relations that merely *constrain* degrees of rational belief.

Another point of disagreement concerns the *uniqueness* of evidential probability, relative to a given state of knowledge. Rudolf Carnap held, for example, that logical principles always determine a unique logical probability for any statement, relative to any body of evidence.Ramsey, by contrast, thought that while degrees of belief are subject to some rational constraints (such as, but not limited to, the axioms of probability) these constraints usually do not determine a unique value. Rational people, in other words, may differ somewhat in their degrees of belief, even if they all have the same information.

Propensity

Propensity theorists think of probability as a physical propensity, or disposition, or tendency of a given type of physical situation to yield an outcome of a certain kind or to yield a long run relative frequency of such an outcome. This kind of objective probability is sometimes called 'chance'.

Propensities, or chances, are not relative frequencies, but purported causes of the observed stable relative frequencies. Propensities are invoked to explain why repeating a certain kind of experiment will generate given outcome types at persistent rates, which are known as propensities or chances. Frequentists are unable to take this approach, since relative frequencies do not exist for single tosses of a coin, but only for large ensembles or collectives (see "single case possible" in the table above). In contrast, a propensitist is able to use the law of large numbers to explain the behaviour of long-run frequencies. This law, which is a consequence of the axioms of probability, says that if (for example) a coin is tossed repeatedly many times, in such a way that its probability of landing heads is the same on each toss, and the outcomes are probabilistically independent, then the relative frequency of heads will be close to the probability of heads on each single toss. This law allows that stable long-run frequencies are a manifestation of invariant *single-case* probabilities. In addition to explaining the emergence of stable relative frequencies, the idea of propensity is motivated by the desire to make sense of single-case probability attributions in quantum mechanics, such as the probability of decay of a particular atom at a particular time.

The main challenge facing propensity theories is to say exactly what propensity means. (And then, of course, to show that propensity thus defined has the required properties.) At present, unfortunately, none of the well-recognised accounts of propensity comes close to meeting this challenge.

A propensity theory of probability was given by Charles Sanders Peirce. A later propensity theory was proposed by philosopher Karl Popper, who had only slight acquaintance with the writings of C. S. Peirce, however. Popper noted that the outcome of a physical experiment is produced by a certain set of "generating conditions". When we repeat an experiment, as the saying goes, we really perform another experiment with a (more or less) similar set of generating conditions. To say that a set of generating conditions has propensity p of producing the outcome E means that those exact conditions, if repeated indefinitely, would produce an outcome sequence in which E occurred with limiting relative frequency p. For Popper then, a deterministic experiment would have propensity 0 or 1 for each outcome, since those generating conditions would have same outcome on each trial. In other words, non-trivial propensities (those that differ from 0 and 1) only exist for genuinely indeterministic experiments.

A number of other philosophers, including David Miller and Donald A. Gillies, have proposed propensity theories somewhat similar to Popper's.

Other propensity theorists (e.g. Ronald Giere) do not explicitly define propensities at all, but rather see propensity as defined by the theoretical role it plays in science. They argue, for example, that physical magnitudes such as electrical charge cannot be explicitly defined either, in terms of more basic things, but only in terms of what they do (such as attracting and repelling other electrical charges). In a similar way, propensity is whatever fills the various roles that physical probability plays in science.

What roles does physical probability play in science? What are its properties? One central property of chance is that, when known, it constrains rational belief to take the same numerical value. David Lewis called this the *Principal Principle*, (3.3 & 3.5) a term that philosophers have mostly adopted. For example, suppose you are certain that a particular biased coin has propensity 0.32 to land heads every time it is tossed. What is then the correct price for a gamble that pays $1 if the coin lands heads, and nothing otherwise? According to the Principal Principle, the fair price is 32 cents.

Subjectivism

Gambling odds reflect the average bettor's 'degree of belief' in the outcome.

Subjectivists, also known as Bayesians or followers of epistemic probability, give the notion of probability a subjective status by regarding it as a measure of the ‹degree of belief' of the individual assessing the uncertainty of a particular situation. Epistemic or subjective probability is sometimes called credence, as opposed to the term chance for a propensity probability.

Some examples of epistemic probability are to assign a probability to the proposition that a proposed law of physics is true, and to determine how probable it is that a suspect committed a crime, based on the evidence presented.

Gambling odds don't reflect the bookies' belief in a likely winner, so much as the other bettors' belief, because the bettors are actually betting against one another. The odds are set based on how many people have bet on a possible winner, so that even if the high odds players always win, the bookies will always make their percentages anyway.

The use of Bayesian probability raises the philosophical debate as to whether it can contribute valid justifications of belief.

Bayesians point to the work of Ramsey (p 182) and de Finetti (p 103) as proving that subjective beliefs must follow the laws of probability if they are to be coherent. Evidence casts doubt that humans will have coherent beliefs.

The use of Bayesian probability involves specifying a prior probability. This may be obtained from consideration of whether the required prior probability is greater or lesser than a reference probability associated with an urn model or a thought experiment. The issue is that for a given problem, multiple thought experiments could apply, and choosing one is a matter of judgement: different people may assign different prior probabilities, known as the reference class problem. The "sunrise problem" provides an example.

Prediction

An alternative account of probability emphasizes the role of *prediction* – predicting future observations on the basis of past observations, not on unobservable parameters. In its modern form, it is mainly in the Bayesian vein. This was the main function of probability before the 20th century, but fell out of favor compared to the parametric approach, which modeled phenomena as a physical system that was observed with error, such as in celestial mechanics. The modern predictive approach was pioneered by Bruno de Finetti, with the central idea of exchangeability – that future observations should behave like past observations. This view came to the attention of the Anglophone world with the 1974 translation of de Finetti's book, and has since been propounded by such statisticians as Seymour Geisser.

Axiomatic Probability

The mathematics of probability can be developed on an entirely axiomatic basis that is independent of any interpretation: see the articles on probability theory and probability axioms for a detailed treatment.

Etymology

The word probability derives from the Latin probabilitas, which can also mean "probity", a measure of the authority of a witness in a legal case in Europe, and often correlated with the witness's nobility. In a sense, this differs much from the modern meaning of probability, which, in contrast, is a measure of the weight of empirical evidence, and is arrived at from inductive reasoning and statistical inference.

The scientific study of probability is a modern development of mathematics. Gambling shows that there has been an interest in quantifying the ideas of probability for millennia, but exact mathematical descriptions arose much later. There are reasons of course, for the slow development of the mathematics of probability. Whereas games of chance provided the impetus for the mathematical study of probability, fundamental issues are still obscured by the superstitions of gamblers.

Christiaan Huygens probably published the first book on probability

According to Richard Jeffrey, "Before the middle of the seventeenth century, the term 'probable' (Latin *probabilis*) meant *approvable*, and was applied in that sense, univocally, to opinion and to action. A probable action or opinion was one such as sensible people would undertake or hold, in the circumstances." However, in legal contexts especially, 'probable' could also apply to propositions for which there was good evidence.

Gerolamo Cardano

The sixteenth century Italian polymath Gerolamo Cardano demonstrated the efficacy of defining odds as the ratio of favourable to unfavourable outcomes (which implies that the probability of an event is given by the ratio of favourable outcomes to the total number of possible outcomes). Aside from the elementary work by Cardano, the doctrine of probabilities dates to the correspondence of Pierre de Fermat and Blaise Pascal (1654). Christiaan Huygens (1657) gave the earliest known scientific treatment of the subject. Jakob Bernoulli's *Ars Conjectandi* (posthumous, 1713) and Abraham de Moivre's *Doctrine of Chances* (1718) treated the subject as a branch of mathematics. See Ian Hacking's *The Emergence of Probability* and James Franklin's *The Science of Conjecture* for histories of the early development of the very concept of mathematical probability.

The theory of errors may be traced back to Roger Cotes's *Opera Miscellanea* (posthumous, 1722), but a memoir prepared by Thomas Simpson in 1755 (printed 1756) first applied the theory to the discussion of errors of observation. The reprint (1757) of this memoir lays down the axioms that positive and negative errors are equally probable, and that certain assignable limits define the range of all errors. Simpson also discusses continuous errors and describes a probability curve.

The first two laws of error that were proposed both originated with Pierre-Simon Laplace. The first law was published in 1774 and stated that the frequency of an error could be expressed as an exponential function of the numerical magnitude of the error, disregarding sign. The second law of error was proposed in 1778 by Laplace and stated that the frequency of the error is an exponential function of the square of the error. The second law of error is called the normal distribution or the Gauss law. "It is difficult historically to attribute that law to Gauss, who in spite of his well-known precocity had probably not made this discovery before he was two years old."

Daniel Bernoulli (1778) introduced the principle of the maximum product of the probabilities of a system of concurrent errors.

Carl Friedrich Gauss

Adrien-Marie Legendre (1805) developed the method of least squares, and introduced it in his *Nouvelles méthodes pour la détermination des orbites des comètes* (*New Methods for Determining the Orbits of Comets*). In ignorance of Legendre's contribution, an Irish-American writer, Robert Adrain, editor of "The Analyst" (1808), first deduced the law of facility of error,

$$\phi(x) = ce^{-h^2x^2},$$

where h is a constant depending on precision of observation, and c is a scale factor ensuring that the area under the curve equals 1. He gave two proofs, the second being essentially the same as John Herschel's (1850).Gauss gave the first proof that seems to have been known in Europe (the third after Adrain's) in 1809. Further proofs were given by Laplace (1810, 1812), Gauss (1823), James Ivory (1825, 1826), Hagen (1837), Friedrich Bessel (1838), W. F. Donkin (1844, 1856), and Morgan Crofton (1870). Other contributors were Ellis (1844), De Morgan (1864), Glaisher (1872), and Giovanni Schiaparelli (1875). Peters's (1856) formula for r, the probable error of a single observation, is well known.

In the nineteenth century authors on the general theory included Laplace, Sylvestre Lacroix (1816), Littrow (1833), Adolphe Quetelet (1853), Richard Dedekind (1860), Helmert (1872), Hermann Laurent (1873), Liagre, Didion, and Karl Pearson. Augustus De Morgan and George Boole improved the exposition of the theory.

Andrey Markov introduced the notion of Markov chains (1906), which played an important role in stochastic processes theory and its applications. The modern theory of probability based on the measure theory was developed by Andrey Kolmogorov (1931).

On the geometric side (see integral geometry) contributors to *The Educational Times* were influential (Miller, Crofton, McColl, Wolstenholme, Watson, and Artemas Martin).

PROBABILITY THEORY

Probability theory is the branch of mathematics concerned with probability, the analysis of random phenomena. The central objects of probability theory are random variables, stochastic processes, and events: mathematical abstractions of non-deterministic events or measured quantities that may either be single occurrences or evolve over time in an apparently random fashion.

It is not possible to predict precisely results of random events. However, if a sequence of individual events, such as coin flipping or the roll of dice, is influenced by other factors, such as friction, it will exhibit certain patterns, which can be studied and predicted. Two representative mathematical results describing such patterns are the law of large numbers and the central limit theorem.

As a mathematical foundation for statistics, probability theory is essential to many human activities that involve quantitative analysis of large sets of data. Methods of probability theory also apply to descriptions of complex systems given only partial knowledge of their state, as in statistical mechanics. A great discovery of twentieth century physics was the probabilistic nature of physical phenomena at atomic scales, described in quantum mechanics.

The mathematical theory of probability has its roots in attempts to analyze games of chance by Gerolamo Cardano in the sixteenth century, and by Pierre de Fermat and Blaise Pascal in the seventeenth century (for example the “problem of points”). Christiaan Huygens published a book on the subject in 1657 and in the 19th century, Pierre Laplace completed what is today considered the classic interpretation.

Initially, probability theory mainly considered **discrete** events, and its methods were mainly combinatorial. Eventually, analytical considerations compelled the incorporation of **continuous** variables into the theory.

This culminated in modern probability theory, on foundations laid by Andrey Nikolaevich Kolmogorov. Kolmogorov combined the notion of sample space, introduced by Richard von Mises, and **measure theory** and presented his axiom system for probability theory in 1933. This became the mostly undisputed axiomatic basis for modern probability theory; but, alternatives exist, such as the adoption of finite rather than countable additivity by Bruno de Finetti.

Treatment

Most introductions to probability theory treat discrete probability distributions and continuous probability distributions separately. The more mathematically advanced measure theory-based treatment of probability covers the discrete, continuous, a mix of the two, and more.

Motivation

Consider an experiment that can produce a number of outcomes. The set of all outcomes is called the *sample space* of the experiment. The *power set* of the sample space (or equivalently, the event space) is formed by considering all different collections of possible results. For example, rolling an honest die produces one of six possible results. One collection of possible results corresponds to getting an odd number. Thus, the subset {1,3,5} is an element of the power set of the sample space of die rolls. These collections are called *events*. In this case, {1,3,5} is the event that the die falls on some odd number. If the results that actually occur fall in a given event, that event is said to have occurred.

Probability is a way of assigning every “event” a value between zero and one, with the requirement that the event made up of all possible results (in our example, the event {1,2,3,4,5,6}) be assigned a value of one. To qualify as a probability distribution, the assignment of values must satisfy the requirement that if you look at a collection of mutually exclusive events (events that contain no common results, e.g., the events {1,6}, {3}, and {2,4} are all mutually exclusive), the probability that one of the events will occur is given by the sum of the probabilities of the individual events.

The probability that any one of the events {1,6}, {3}, or {2,4} will occur is 5/6. This is the same as saying that the probability of event {1,2,3,4,6} is 5/6. This event encompasses the possibility of any number except five being rolled. The mutually exclusive event {5} has a probability of 1/6, and the event {1,2,3,4,5,6} has a probability of 1, that is, absolute certainty.

Discrete Probability Distributions

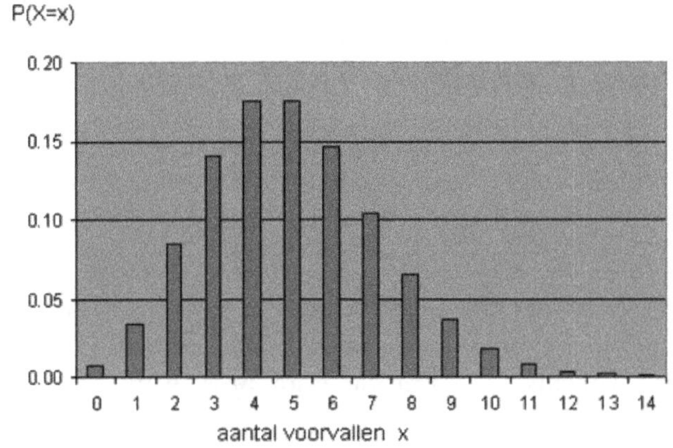

The Poisson distribution, a discrete probability distribution.

Discrete probability theory deals with events that occur in countable sample spaces.

Examples: Throwing dice, experiments with decks of cards, random walk, and tossing coins

Classical definition: Initially the probability of an event to occur was defined as number of cases favorable for the event, over the number of total outcomes possible in an equiprobable sample space: For example, if the event is "occurrence of an even number when a die is rolled", the probability is given by $\frac{3}{6} = \frac{1}{2}$, since 3 faces out of the 6 have even numbers and each face has the same probability of appearing.

Modern definition: The modern definition starts with a finite or countable set called the **sample space**, which relates to the set of all *possible outcomes* in classical sense, denoted by Ω It is then assumed that for each element $x \in \Omega$ an intrinsic "probability" value f(x), is attached, which satisfies the following properties:

1. $f(x) \in [0,1]$ for all $x \in \Omega$;
2. $\sum_{x\in\Omega} f(x) = 1$.

That is, the probability function f(x) lies between zero and one for every value of x in the sample space Ω, and the sum of f(x) over all values x in the sample space Ω is equal to 1. An event is defined as any subset E , of the sample space Ω,. The probability of the event E, is defined as

$$P(E) = \sum_{x\in E} f(x).$$

So, the probability of the entire sample space is 1, and the probability of the null event is 0.

The function $f(x)$ mapping a point in the sample space to the "probability" value is called a **probability mass function** abbreviated as **pmf**. The modern definition does not try to answer how probability mass functions are obtained; instead it builds a theory that assumes their existence

Measure-Theoretic Probability Theory

The *raison d'être* of the measure-theoretic treatment of probability is that it unifies the discrete and the continuous cases, and makes the difference a question of which measure is used. Furthermore, it covers distributions that are neither discrete nor continuous nor mixtures of the two.

An example of such distributions could be a mix of discrete and continuous distributions—for example, a random variable that is 0 with probability 1/2, and takes a random value from a normal distribution with probability 1/2. It can still be studied to some extent by considering it to have a pdf of of $(\delta[x] + \varphi(x))/2$ where δ is the Dirac delta function.

Other distributions may not even be a mix, for example, the Cantor distribution has no positive probability for any single point, neither does it have a density. The modern approach to probability theory solves these problems using measure theory to define the probability space:

Given any set Ω, (also called sample space) and a σ-algebra F, on it, a measure P, defined on F, is called a probability measure if $P(\Omega) = 1$. If F, is the Borel σ-algebra on the set of real numbers, then there is a unique probability measure on F, for any cdf, and vice versa. The measure corresponding to a cdf is said to be induced by the cdf. This measure coincides with the pmf for discrete variables and pdf for continuous variables, making the measure-theoretic approach free of fallacies.

The probability of a set E , in the σ-algebra F, is defined as

$$P(E) = \int_{\omega \in E} \mu_F(d\omega)$$

where the integration is with respect to the measure μ F, induced by F.

Along with providing better understanding and unification of discrete and continuous probabilities, measure-theoretic treatment also allows us to work on probabilities outside $\mathbb{R}^n$, as in the theory of stochastic processes. For example, to study Brownian motion, probability is defined on a space of functions.

When it's convenient to work with a dominating measure, the Radon-Nikodym theorem is used to define a density as the Radon-Nikodym derivative of the probability distribution of interest with respect to this dominating measure. Discrete densities are usually defined as this derivative with respect to a counting measure over the set of all possible outcomes. Densities for absolutely continuous distributions are usually defined as this derivative with respect to the Lebesgue measure. If a theorem can be proved in this general setting, it holds for both discrete and continuous distributions as well as others; separate proofs are not required for discrete and continuous distributions.

Classical Probability Distributions

Certain random variables occur very often in probability theory because they well describe many natural or physical processes. Their distributions therefore have gained *special importance* in probability theory. Some fundamental *discrete distributions* are the discrete uniform, Bernoulli, binomial, negative binomial, Poisson and geometric distributions. Important *continuous distributions* include the continuous uniform, normal, exponential, gamma and beta distributions.

Convergence of Random Variables

In probability theory, there are several notions of convergence for random variables. They are listed below in the order of strength, i.e., any subsequent notion of convergence in the list implies convergence according to all of the preceding notions.

Law of Large Numbers

Common intuition suggests that if a fair coin is tossed many times, then roughly half of the time it will turn up heads, and the other half it will turn up tails. Furthermore, the more often the coin is tossed, the more likely it should be that the ratio of the number of heads to the number of tails will approach unity. Modern probability theory provides a formal version of this intuitive idea, known as the law of large numbers. This law is remarkable because it is not assumed in the foundations of probability theory, but instead emerges from these foundations as a theorem. Since it links theoretically derived probabilities to their actual frequency of occurrence in the real world, the law of large numbers is considered as a pillar in the history of statistical theory and has had widespread influence.

The law of large numbers (LLN) states that the sample average

$$\overline{X}_n = \frac{1}{n}\sum_{k=1}^{n} X_k$$

of a sequence of independent and identically distributed random variables X_k converges towards their common expectation μ , provided that the expectation of $|X_k|$ is finite.

It is in the different forms of convergence of random variables that separates the weak and the strong law of large numbers

Weak law: $\overline{X}_n \xrightarrow{P} \mu$ for $n \to \infty$

Strong law: $\overline{X}_n \xrightarrow{\text{a. s.}} \mu$ for $n \to \infty$.

It follows from the LLN that if an event of probability p is observed repeatedly during independent experiments, the ratio of the observed frequency of that event to the total number of repetitions converges towards p.

For example, if Y_1, Y_2, . . .are independent Bernoulli random variables taking values 1 with probability p and 0 with probability 1-p, then E (Yi) = p for all i, so that $\bar{Y}$ n converges to p almost surely.

Central Limit Theorem

"The central limit theorem (CLT) is one of the great results of mathematics." (Chapter 18 in) It explains the ubiquitous occurrence of the normal distribution in nature.

The theorem states that the average of many independent and identically distributed random variables with finite variance tends towards a normal distribution *irrespective* of the distribution followed by the original random variables. Formally, let X_1, X_2,... be independent random variables with mean u and variance $\sigma_2 > 0$. Then the sequence of random variables

$$Z_n = \frac{\sum_{i=1}^{n}(X_i - \mu)}{\sigma\sqrt{n}}$$

converges in distribution to a standard normal random variable.

For some classes of random variables the classic central limit theorem works rather fast (see Berry–Esseen theorem), for example the distributions with finite first, second and third moment from the exponential family; on the other hand, for some random variables of the heavy tail and fat tail variety, it works very slowly or may not work at all: in such cases one may use the Generalized Central Limit Theorem (GCLT).

Applications

Probability theory is applied in everyday life in risk assessment and modeling. The insurance industry and markets use actuarial science to determine pricing and make trading decisions. Governments apply probabilistic methods in environmental regulation, entitlement analysis (Reliability theory of aging and longevity), and financial regulation.

A good example of the use of probability theory in equity trading is the effect of the perceived probability of any widespread Middle East conflict on oil prices, which have ripple effects in the economy as a whole. An assessment by a commodity trader that a war is more likely can send that commodity's prices up or down, and signals other traders of that opinion. Accordingly, the probabilities are neither assessed independently nor necessarily very rationally. The theory of behavioral finance emerged to describe the effect of such groupthink on pricing, on policy, and on peace and conflict.

In addition to financial assessment, probability can be used to analyze trends in biology (e.g. disease spread) as well as ecology (e.g. biological Punnett squares). As with finance, risk assessment can be used as a statistical tool to calculate the likelihood of undesirable events occurring and can assist with implementing protocols to avoid encountering such circumstances. Probability is used to design games of chance so that casinos can make a guaranteed profit, yet provide payouts to players that are frequent enough to encourage continued play. The discovery of rigorous methods to assess and combine probability assessments has changed society. It is important for most citizens to understand how probability assessments are made, and how they contribute to decisions.

Another significant application of probability theory in everyday life is reliability. Many consumer products, such as automobiles and consumer electronics, use reliability theory in product design to reduce the probability of failure. Failure probability may influence a manufacturer's decisions on a product's warranty.

The cache language model and other statistical language models that are used in natural language processing are also examples of applications of probability theory.

Mathematical Treatment

Consider an experiment that can produce a number of results. The collection of all possible results is called the sample space of the experiment. The power set of the sample space is formed by considering all different collections of possible results. For example, rolling a dice can produce six possible results. One collection of possible results gives an odd number on the dice. Thus, the subset $\{1,3,5\}$ is an element of the power set of the sample space of dice rolls. These collections are called "events." In this case, $\{1,3,5\}$ is the event that the dice falls on some odd number. If the results that actually occur fall in a given event, the event is said to have occurred.

A probability is a way of assigning every event a value between zero and one, with the requirement that the event made up of all possible results (in our example, the event $\{1,2,3,4,5,6\}$) is assigned a value of one. To qualify as a probability, the assignment of values must satisfy the requirement that if you look at a collection of mutually exclusive events (events with no common results, e.g., the events $\{1,6\}$, $\{3\}$, and $\{2,4\}$ are all mutually exclusive), the probability that at least one of the events will occur is given by the sum of the probabilities of all the individual events.

The probability of an event *A* is written as $P(A)$, $p(A)$, or $\Pr(A)$. This mathematical definition of probability can extend to infinite sample spaces, and even uncountable sample spaces, using the concept of a measure.

The *opposite* or *complement* of an event *A* is the event [not *A*] (that is, the event of *A* not occurring), often denoted as $\overline{A}, A^C, \neg A$, or $\sim A$; its probability is given by *P*(not *A*) = 1 − *P*(*A*). As an example, the chance of not rolling a six on a six-sided die is 1 – (chance of rolling a six) $= 1 - \frac{1}{6} = \frac{5}{6}$.

If two events *A* and *B* occur on a single performance of an experiment, this is called the intersection or joint probability of *A* and *B*, denoted as $P(A \cap B)$.

Independent Events

If two events, *A* and *B* are independent then the joint probability is

$$P(A \text{ and } B) = P(A \cap B) = P(A)P(B),$$

for example, if two coins are flipped the chance of both being heads is $\frac{1}{2} \times \frac{1}{2} = \frac{1}{4}$.

Mutually Exclusive Events

If either event A or event B occurs on a single performance of an experiment this is called the union of the events A and B denoted as $P(A \cup B)$ If two events are mutually exclusive then the probability of either occurring is

$P(A \text{ or } B) = P(A \cup B) = P(A) + P(B)$.

For example, the chance of rolling a 1 or 2 on a six-sided die is $P(1 \text{ or } 2) = P(1) + P(2) = \frac{1}{6} + \frac{1}{6} = \frac{1}{3}$.

Not Mutually Exclusive Events

If the events are not mutually exclusive then

$P(A \text{ or } B) = P(A) + P(B) - P(A \text{ and } B)$.

For example, when drawing a single card at random from a regular deck of cards, the chance of getting a heart or a face card (J,Q,K) (or one that is both) is $\frac{13}{52} + \frac{12}{52} - \frac{3}{52} = \frac{11}{26}$ because of the 52 cards of a deck 13 are hearts, 12 are face cards, and 3 are both: here the possibilities included in the "3 that are both" are included in each of the "13 hearts" and the "12 face cards" but should only be counted once.

Conditional Probability

Conditional probability is the probability of some event A, given the occurrence of some other event B. Conditional probability is written $P(A \mid B)$ and is read "the probability of A, given B". It is defined by

$$P(A \mid B) = \frac{P(A \cap B)}{P(B)}.$$

If P(B)=0 then P (A|B) is is formally undefined by this expression. However, it is possible to define a conditional probability for some zero-probability events using a σ-algebra of such events (such as those arising from a continuous random variable).

For example, in a bag of 2 red balls and 2 blue balls (4 balls in total), the probability of taking a red ball is 1/2 however, when taking a second ball, the probability of it being either a red ball or a blue ball depends on the ball previously taken, such as, if a red ball was taken, the probability of picking a red ball again would be 1/3 since only 1 red and 2 blue balls would have been remaining.

Inverse Probability

In probability theory and applications, Bayes' rule relates the odds of event A_1 to event A_2, before (prior to) and after (posterior to) conditioning on another event B. The odds on A_1 to event A_2 is simply the ratio of the probabilities of the two events. When arbitrarily many events A are of interest, not just two, the rule can be rephrased as posterior is proportional to prior times likelihood, P(A)P(B|A)}∝ P(A|B)\propto P(A)P(B|A) where the proportionality symbol means that the left hand side is proportional to (i.e., equals a constant times) the right hand side as Avaries, for fixed or given B (Lee, 2012; Bertsch McGrayne, 2012). In this form it goes back to Laplace (1774) and to Cournot (1843); see Fienberg (2005). See Inverse probability and Bayes' rule.

Summary of Probabilities

Event	Probability
A	$P(A) \in [0,1]$
not A	$P(A^C) = 1 - P(A)$
A or B	$P(A \cup B) = P(A) + P(B) - P(A \cap B)$ $P(A \cup B) = P(A) + P(B)$ if A and B are mutually exclusive
A and B	$P(A \cap B) = P(A\|B)P(B) = P(B\|A)P(A)$ $P(A \cap B) = P(A)P(B)$ if A and B are independent
A given B	$P(A \mid B) = \frac{P(A \cap B)}{P(B)} = \frac{P(B\|A)P(A)}{P(B)}$

Relation to Randomness

In a deterministic universe, based on Newtonian concepts, there would be no probability if all conditions were known (Laplace's demon), (but there are situations in which sensitivity to initial conditions exceeds our ability to measure them, i.e. know them). In the case of a roulette wheel, if the force of the hand and the period of that force are known, the number on which the ball will stop would be a certainty (though as a practical matter, this would likely be true only of a roulette wheel that had not been exactly levelled — as Thomas A. Bass' Newtonian Casino revealed). Of course, this also assumes knowledge of inertia and friction of the wheel, weight, smoothness and roundness of the ball, variations in hand speed during the turning and so forth. A probabilistic description can thus be more useful than Newtonian mechanics for analyzing the pattern of outcomes of repeated rolls of a roulette wheel. Physicists face the same situation in kinetic theory of gases, where the system, while deterministic *in principle*, is so complex (with the number of molecules typically the order of magnitude of Avogadro constant $6.02 \cdot 10^{23}$) that only a statistical description of its properties is feasible.

Probability theory is required to describe quantum phenomena. A revolutionary discovery of early 20th century physics was the random character of all physical processes that occur at sub-atomic scales and are governed by the laws of quantum mechanics. The objective wave function evolves deterministically but, according to the Copenhagen interpretation, it deals with probabilities of observing, the outcome being explained by a wave function collapse when an observation is made. However, the loss of determinism for the sake of instrumentalism did not meet with universal approval. Albert Einstein famously remarked in a letter to Max Born: "I am convinced that God does not play dice". Like Einstein, Erwin Schrödinger, who discovered the wave function, believed quantum mechanics is a statistical approximation of an underlying deterministic reality. In some modern interpretations of the statistical mechanics of measurement, quantum decoherence is invoked to account for the appearance of subjectively probabilistic experimental outcomes.

REFERENCES

1. "Probability". Webster's Revised Unabridged Dictionary. G & C Merriam, 1913
2. Strictly speaking, a value of 0 indicates impossibility only when a sample space is finite. In an infinite sample space, a probability of 0 may indicate infinitesimal likelihood, but not impossibility. For example, the probability of any specific integer, say 5, being chosen completely at random from the entire set of integers is 0. However, it is not impossible for 5 to be chosen, just extremely unlikely.
3. "Kendall's Advanced Theory of Statistics, Volume 1: Distribution Theory", Alan Stuart and Keith Ord, 6th Ed, (2009), ISBN 9780534243128
4. William Feller, "An Introduction to Probability Theory and Its Applications", (Vol 1), 3rd Ed, (1968), Wiley, ISBN 0-471-25708-7
5. Probability Theory The Britannica website
6. Hacking, Ian (1965). The Logic of Statistical Inference. Cambridge University Press. ISBN 0-521-05165-7.[page needed]
7. Finetti, Bruno de (1970). "Logical foundations and measurement of subjective probability". Acta Psychologica. 34: 129–145. doi:10.1016/0001-6918(70)90012-0.
8. Hájek, Alan. "Interpretations of Probability". The Stanford Encyclopedia of Philosophy (Winter 2012 Edition), Edward N. Zalta (ed.). Retrieved 22 April 2013.
9. Hogg, Robert V.; Craig, Allen; McKean, Joseph W. (2004). Introduction to Mathematical Statistics (6th ed.). Upper Saddle River: Pearson. ISBN 0-13-008507-3.[page needed]
10. Hacking, I. (2006) The Emergence of Probability: A Philosophical Study of Early Ideas about Probability, Induction and Statistical Inference, Cambridge University Press, ISBN 978-0-521-68557-3[page needed]
11. Freund, John. (1973) Introduction to Probability. Dickenson ISBN 978-0822100782 (p. 1)
12. Jeffrey, R.C., Probability and the Art of Judgment, Cambridge University Press. (1992). pp. 54-55 . ISBN 0-521-39459-7

13. Franklin, J. (2001) The Science of Conjecture: Evidence and Probability Before Pascal, Johns Hopkins University Press. (pp. 22, 113, 127)
14. Some laws and problems in classical probability and how Cardano anticipated them Gorrochum, P. Chance magazine 2012
15. Abrams, William, A Brief History of Probability, Second Moment, retrieved 2008-05-23
16. Ivancevic, Vladimir G.; Ivancevic, Tijana T. (2008). Quantum leap : from Dirac and Feynman, across the universe, to human body and mind. Singapore ; Hackensack, NJ: World Scientific. p. 16. ISBN 978-981-281-927-7.
17. Franklin, James (2001). The Science of Conjecture: Evidence and Probability Before Pascal. Johns Hopkins University Press. ISBN 0801865697.
18. Wilson EB (1923) "First and second laws of error". Journal of the American Statistical Association, 18, 143
19. Seneta, Eugene William. ""Adrien-Marie Legendre" (version 9)". StatProb: The Encyclopedia Sponsored by Statistics and Probability Societies. Retrieved 27 January 2016.
20. http://www.statslab.cam.ac.uk/~rrw1/markov/M.pdf
21. Vitanyi, Paul M.B. (1988). "Andrei Nikolaevich Kolmogorov". CWI Quarterly (1): 3–18. Retrieved 27 January 2016.
22. Singh, Laurie (2010) "Whither Efficient Markets? Efficient Market Theory and Behavioral Finance". The Finance Professionals' Post, 2010.
23. Gao, J.Z.; Fong, D.; Liu, X. (April 2011). "Mathematical analyses of casino rebate systems for VIP gambling". International Gambling Studies. 11 (1): 93–106. doi:10.1080/14459795.2011.552575.
24. Gorman, Michael (2011) "Management Insights". Management Science[full citation needed]
25. Ross, Sheldon. A First course in Probability, 8th Edition. Page 26-27.
26. Olofsson (2005) Page 8.
27. Olofsson (2005), page 9
28. Olofsson (2005) page 35.
29. Olofsson (2005) page 29.
30. Burgi, Mark (2010) "Interpretations of Negative Probabilities", p. 1. arXiv:1008.1287v1
31. Jedenfalls bin ich überzeugt, daß der Alte nicht würfelt. Letter to Max Born, 4 December 1926, in: Einstein/Born Briefwechsel 1916-1955.
32. Moore, W.J. (1992). Schrödinger: Life and Thought. Cambridge University Press. p. 479. ISBN 0-521-43767-9.
33. de Elía, Ramón; Laprise, René (2005). "Diversity in interpretations of probability: implications for weather forecasting". Monthly Weather Review. 133 (5): 1129–1143. doi:10.1175/mwr2913.1. "There are several schools of thought regarding the

interpretation of probabilities, none of them without flaws, internal contradictions, or paradoxes." (p 1129) "There are no standard classifications of probability interpretations, and even the more popular ones may suffer subtle variations from text to text." (p 1130) The classification in this article is representative, as are the authors and ideas claimed for each classification.

34. Venn, John (1876). The Logic of Chance. London: MacMillan. Retrieved August 2013. Check date values in: |access-date= (help)
35. Reichenbach, Hans (1948). The theory of probability, an inquiry into the logical and mathematical foundations of the calculus of probability. University of California Press. English translation of the original 1935 German. ASIN: B000R0D5MS
36. Mises, Richard (1981). Probability, statistics, and truth. New York: Dover Publications. ISBN 0-486-24214-5. English translation of the third German edition of 1951 which was published 30 years after the first German edition.
37. Rowbottom, Darrell (2015). Probability. Cambridge: Polity. ISBN 978-0745652573.
38. Laplace, P. S., 1814, English edition 1951, A Philosophical Essay on Probabilities, New York: Dover Publications Inc.
39. (1964). "Foresight: its Logical laws, its Subjective Sources". In Kyburg, H. E. Studies in Subjective Probability. H. E. Smokler. New York: Wiley. pp. 93–158. Translation of the 1937 French original with later notes added.
40. Savage, L.J. (1954). The foundations of statistics. New York: John Wiley & Sons, Inc. ISBN 0-486-62349-1.
41. ^ Jump up to:a b c Ramsey, F. P. (1931). "Chapter VII, Truth and Probability (1926)". In Braithwaite, R. B. Foundations of Mathematics and Other Logical Essays (PDF). London: Kegan, Paul, Trench, Trubner & Co. pp. 156–198. Retrieved August 2013. Check date values in: |access-date=(help) Contains three chapters (essays) by Ramsey. The electronic version contains only those three.
42. Cox, Richard Threlkeld (1961). The algebra of probable inference. Baltimore: Johns Hopkins Press.
43. Keynes, John Maynard (1921). A Treatise on Probability. MacMillan. Retrieved August 2013. Check date values in: |access-date= (help)
44. Carnap, Rudolph (1950). Logical Foundations of Probability. Chicago: University of Chicago Press. Carnap coined the notion "probability1" and "probability2" for evidential and physical probability, respectively.
45. Gillies, Donald (2000). Philosophical theories of probability. London New York: Routledge. ISBN 978-0415182768.
46. Fermat and Pascal on Probability (@ socsci.uci.edu)
47. Laszlo E. Szabo, A Physicalist Interpretation of Probability (Talk presented on the Philosophy of Science Seminar, Eötvös, Budapest, 8 October 2001.)
48. Laszlo E. Szabo, Objective probability-like things with and without objective indeterminism, Studies in History and Philosophy of Modern Physics 38 (2007)

626–634 (Preprint)

49. Spanos, Aris (1986). Statistical foundations of econometric modelling. Cambridge New York: Cambridge University Press. ISBN 978-0521269124.
50. Freedman, David and Philip B. Stark (2003)"What is the Chance of an Earthquake?" Earthquake Science and Seismic Risk.
51. Peterson, Martin (2009). An introduction to decision theory. Cambridge, UK New York: Cambridge University Press. p. 140. ISBN 978-0521716543.
52. Miller, Richard W. (1975). "Propensity: Popper or Peirce?". British Journal for the Philosophy of Science. 26 (2): 123–132. doi:10.1093/bjps/26.2.123.
53. Haack, Susan; Kolenda, Konstantin, Konstantin; Kolenda (1977). "Two Fallibilists in Search of the Truth". Proceedings of the Aristotelian Society. 51 (Supplementary Volumes): 63–104. JSTOR 4106816.
54. Burks, Arthur W. (1978). Chance, Cause and Reason: An Inquiry into the Nature of Scientific Evidence. University of Chicago Press. pp. 694 pages. ISBN 0-226-08087-0.
55. Peirce, Charles Sanders and Burks, Arthur W., ed. (1958), the Collected Papers of Charles Sanders Peirce Volumes 7 and 8, Harvard University Press, Cambridge, MA, also Belnap Press (of Harvard University Press) edition, vols. 7-8 bound together, 798 pages, online via InteLex, reprinted in 1998 Thoemmes Continuum.
56. Ronald N. Giere (1973). "Objective Single Case Probabilities and the Foundations of Statistics". Studies in Logic and the Foundations of Mathematics. 73. Elsevier. ISBN 978-0-444-10491-5.
57. Jaynes, E. T. (2003). Probability theory the logic of science. Cambridge, UK New York, NY: Cambridge University Press. ISBN 978-0521592710.
58. Kahneman, Daniel (2011). Thinking, fast and slow. New York: Farrar, Straus and Giroux. ISBN 978-0374275631. The book contains numerous examples of the difference between idealized and actual thought. "[W]hen called upon to judge probability, people actually judge something else and believe they have judged probability." (p 98)
59. Grove, William M.; Meehl, Paul E. (1996). "Comparative efficiency of informal (subjective, impressionistic) and formal (mechanical, algorithmic) prediction procedures: The clinical-statistical controversy". Psychology, Public Policy, and Law. 2 (2): 293–332. doi:10.1037/1076-8971.2.2.293. Statistical decisions are consistently superior to the subjective decisions of experts.
60. ^ Jump up to:a b c Predictive Inference: An Introduction, Seymour Geisser, CRC Press, 1993 ISBN 0-412-03471-9
61. Bayes
62. Dice Rolls are Not Completely Random | Inside Science
63. Inferring From Data
64. "Why is quantum mechanics based on probability theory?". StackExchange. July 1, 2014.[unreliable source?]

65. Grinstead, Charles Miller; James Laurie Snell. "Introduction". Introduction to Probability. pp. vii.

66. Hájek, Alan. "Interpretations of Probability". Retrieved 2012-06-20.

67. ""The origins and legacy of Kolmogorov's Grundbegriffe", by Glenn Shafer and Vladimir Vovk" (PDF). Retrieved 2012-02-12.

68. Ross, Sheldon (2010). A First Course in Probability (8th ed.). Pearson Prentice Hall. pp. 26–27. ISBN 978-0-13-603313-4. Retrieved 2016-02-28.

69. "Leithner & Co Pty Ltd - Value Investing, Risk and Risk Management - Part I". Leithner.com.au. 2000-09-15. Archived from the original on 2014-01-26. Retrieved 2012-02-12.

70. David Williams, "Probability with martingales", Cambridge 1991/2008

CHAPTER 8

Real Analysis

Chapter Contents

The subject of real analysis is concerned with studying the behavior and properties of functions, sequences, and sets on the real number line, which we denote as the mathematically familiar **R**. Concepts that we wish to examine through real analysis include properties like Limits, Continuity, Derivatives (rates of change), and Integration (amount of change over time). Many of these ideas are, on a conceptual or practical level, dealt with at lower levels of mathematics, including a regular First-Year Calculus course, and so, to the uninitiated reader, the subject of Real Analysis may seem rather senseless and trivial. However, Real Analysis is at a depth, complexity, and arguably beauty, that it is because under the surface of everyday mathematics, there is an assurance of correctness, that we call rigor, that permeates the whole of mathematics. Thus, Real Analysis can, to some degree, be viewed as a development of a rigorous, well-proven framework to support the intuitive ideas that we frequently take for granted.

INTRODUCTION

Real analysis (traditionally, the theory of functions of a real variable) is a branch of mathematical analysis dealing with the real numbers and real-valued functions of a real variable. In particular, it deals with the analytic properties of real functions and sequences, including convergence and limits of sequences of real numbers, the calculus of the real numbers, and continuity, smoothness and related properties of real-valued functions.

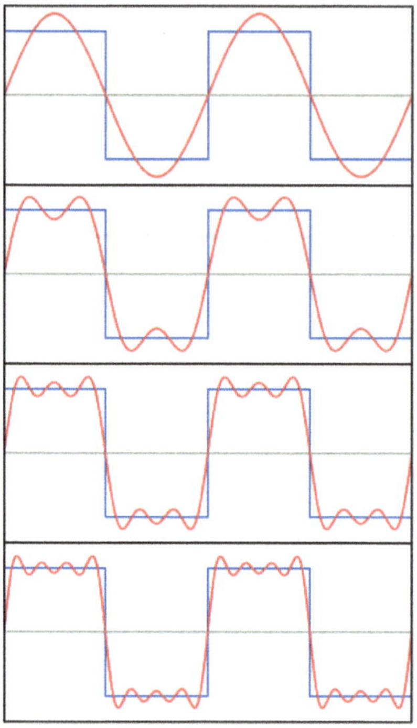

The first four partial sums of the Fourier series for a square wave. Fourier series are an important tool in real analysis.

CONSTRUCTION OF THE REAL NUMBERS

In mathematics, there are several ways of defining the real number system as an ordered field. The synthetic approach gives a list of axioms for the real numbers as a complete ordered field. Under the usual axioms of set theory, one can show that these axioms are categorical, in the sense that there is a model for the axioms, and any two such models are isomorphic. Any one of these models must be explicitly constructed, and most of these models are built using the basic properties of the rational number system as an ordered field.

Synthetic Approach

The synthetic approach axiomatically defines the real number system as a complete ordered field. Precisely, this means the following. A model for the real number system consists of a set **R**, two distinct elements 0 and 1 of **R**, two binary operations + and × on **R** (called addition and multiplication, respectively), and a binary relation ≤ on **R**, satisfying the following properties.

- (**R**, +, ×) forms a field. In other words,
  - For all x, y, and z in **R**, $x + (y + z) = (x + y) + z$ and $x \times (y \times z) = (x \times y) \times z$. (associativity of addition and multiplication)
  - For all x and y in **R**, $x + y = y + x$ and $x \times y = y \times x$. (commutativity of addition and multiplication)
  - For all x, y, and z in **R**, $x \times (y + z) = (x \times y) + (x \times z)$. (distributivity of multiplication over addition)
  - For all x in **R**, $x + 0 = x$. (existence of additive identity)
  - 0 is not equal to 1, and for all x in **R**, $x \times 1 = x$. (existence of multiplicative identity)
  - For every x in **R**, there exists an element $-x$ in **R**, such that $x + (-x) = 0$. (existence of additive inverses)
  - For every $x \neq 0$ in **R**, there exists an element x^{-1} in **R**, such that $x \times x^{-1} = 1$. (existence of multiplicative inverses)
- (**R**, ≤) forms a totally ordered set. In other words,
  - For all x in **R**, $x \leq x$. (reflexivity)
  - For all x and y in **R**, if $x \leq y$ and $y \leq x$, then $x = y$. (antisymmetry)
  - For all x, y, and z in **R**, if $x \leq y$ and $y \leq z$, then $x \leq z$. (transitivity)
  - For all x and y in **R**, $x \leq y$ or $y \leq x$. (totalness)
- The field operations + and × on **R** are compatible with the order ≤. In other words,
  - For all x, y and z in **R**, if $x \leq y$, then $x + z \leq y + z$. (preservation of order under addition)
  - For all x and y in **R**, if $0 \leq x$ and $0 \leq y$, then $0 \leq x \times y$ (preservation of order under multiplication)
- The order ≤ is complete in the following sense: every non-empty subset of **R** bounded above has a least upper bound. In other words,
  - If A is a non-empty subset of **R**, and if A has an upper bound, then A has a least upper bound u, such that for every upper bound v of A, $u \leq v$.

The rational numbers **Q** satisfy the first three axioms (i.e. **Q** is totally ordered field) but **Q** does not satisfy axiom 4. So axiom 4, which requires the order to be Dedekind-complete, is crucial. Axiom 4 implies the Archimedean property. Several models for axioms 1-4 are given below. Any two models for axioms 1-4 are isomorphic, and so up to isomorphism, there is only one complete ordered Archimedean field.

When we say that any two models of the above axioms are isomorphic, we mean that for any two models $(R, 0_R, 1_R, +_R, \times_R, \leq_R)$ and $(S, 0_S, 1_S, +_S, \times_S, \leq_S)$, there is a bijection $f : R \rightarrow S$ preserving both the field operations and the order. Explicitly,

- f is both injective and surjective.
- $f(0_R) = 0_S$ and $f(1_R) = 1_S$.

- For all x and y in R, $f(x +_R y) = f(x) +_S f(y)$ and $f(x \times_R y) = f(x) \times_S f(y)$.
- For all x and y in R, $x \leq_R y$ if and only if $f(x) \leq_S f(y)$.

Explicit Constructions of Models

We shall not prove that any models of the axioms are isomorphic. Such a proof can be found in any number of modern analysis or set theory textbooks. We will sketch the basic definitions and properties of a number of constructions, however, because each of these is important for both mathematical and historical reasons. The first three, due to Georg Cantor/Charles Méray, Richard Dedekind and Karl Weierstrass/Otto Stolz all occurred within a few years of each other. Each has advantages and disadvantages. A major motivation in all three cases was the instruction of mathematics students.

Construction from Cauchy Sequences

A standard procedure to force all Cauchy sequences in a metric space to converge is adding new points to the metric space in a process called completion.

R is defined as the completion of **Q** with respect to the metric |x-y|, as will be detailed below (for completions of **Q** with respect to other metrics, see p-adic numbers.)

Let R be the set of Cauchy sequences of rational numbers. That is, sequences

$x_1, x_2, x_3,\ldots$

of rational numbers such that for every rational $\varepsilon > 0$, there exists an integer N such that for all natural numbers $m,n > N$, $|x_m - x_n| < \varepsilon$. Here the vertical bars denote the absolute value.

Cauchy sequences (x_n) and (y_n) can be added and multiplied as follows:

$(x_n) + (y_n) = (x_n + y_n)$

$(x_n) \times (y_n) = (x_n \times y_n)$.

Two Cauchy sequences are called equivalent if and only if the difference between them tends to zero. This defines an equivalence relation that is compatible with the operations defined above, and the set **R** of all equivalence classes can be shown to satisfy all axioms of the real numbers. We can embed **Q** into **R** by identifying the rational number r with the equivalence class of the sequence (r,r,r, ...).

Comparison between real numbers is obtained by defining the following comparison between Cauchy sequences: $(x_n) \geq (y_n)$ if and only if x is equivalent to y or there exists an integer N such that $x_n \geq y_n$ for all $n > N$.

By construction, every real number x is represented by a Cauchy sequence of rational numbers. This representation is far from unique; every rational sequence that converges to x is a representation of x. This reflects the observation that one can often use different sequences to approximate the same real number.

The only real number axiom that does not follow easily from the definitions is the completeness of ≤, i.e. the least upper bound property. It can be proved as follows: Let S be a non-empty subset of **R** and U be an upper bound for S. Substituting a larger value if necessary, we may assume U is rational. Since S is non-empty, we can choose a rational number L such that L < s for some s in S. Now define sequences of rationals (u_n) and (l_n) as follows:

Set $u_0 = U$ and $l_0 = L$.

For each n consider the number:

$m_n = (u_n + l_n)/2$

If m_n is an upper bound for S set:

$u_{n+1} = m_n$ and $l_{n+1} = l_n$

Otherwise set:

$l_{n+1} = m_n$ and $u_{n+1} = u_n$

This obviously defines two Cauchy sequences of rationals, and so we have real numbers $l = (l_n)$ and $u = (u_n)$. It is easy to prove, by induction on n that:

u_n is an upper bound for S for all n

and:

l_n is never an upper bound for S for any n

Thus u is an upper bound for S. To see that it is a least upper bound, notice that the limit of $(u_n - l_n)$ is 0, and so $l = u$. Now suppose $b < u = l$ is a smaller upper bound for S. Since (l_n) is monotonic increasing it is easy to see that $b < l_n$ for some n. But l_n is not an upper bound for S and so neither is b. Hence u is a least upper bound for S and $\leq$ is complete.

The usual decimal notation can be translated to Cauchy sequences in a natural way. For example, the notation $\pi = 3.1415...$ means that π is the equivalence class of the Cauchy sequence (3, 3.1, 3.14, 3.141, 3.1415, ...). The equation 0.999... = 1 states that the sequences (0, 0.9, 0.99, 0.999,...) and (1, 1, 1, 1,...) are equivalent, i.e., their difference converges to 0.

An advantage of constructing **R** as the completion of **Q** is that this construction is not specific to one example; it is used for other metric spaces as well.

Construction by Dedekind Cuts

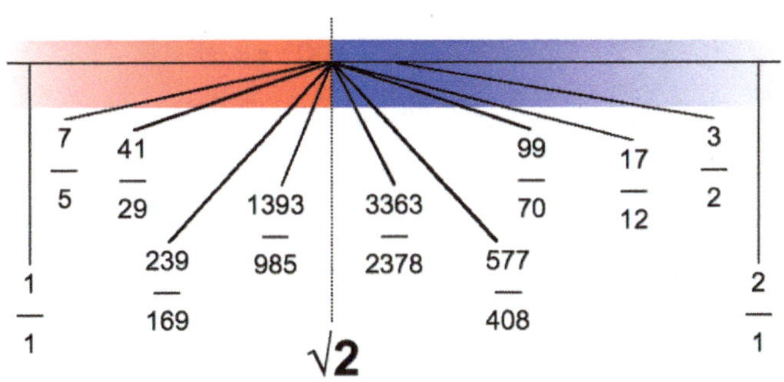

Dedekind used his cut to construct the irrational, real numbers.

A Dedekind cut in an ordered field is a partition of it, (A, B), such that A is nonempty and closed downwards, B is nonempty and closed upwards, and A contains no greatest element. Real numbers can be constructed as Dedekind cuts of rational numbers.

Construction by Dedekind Cuts

A Dedekind cut in an ordered field is a partition of it, (A, B), such that A is nonempty and closed downwards, B is nonempty and closed upwards, and A contains no greatest element. Real numbers can be constructed as Dedekind cuts of rational numbers.

For convenience we may take the lower set A, as the representative of any given Dedekind cut (A,B), since A completely determines B . By doing this we may think intuitively of a real number as being represented by the set of all smaller rational numbers. In more detail, a real number r is any subset of the set Q of rational numbers that fulfills the following conditions:

1. r is not empty
2. $r \neq \mathbf{Q}$
3. r is closed downwards. In other words, for all $x, y \in \mathbf{Q}$ such that $x < y$, if $y \in r$ then $x \in r$
4. r contains no greatest element. In other words, there is no $x \in r$ such that for all $y \in r$, $y \leq x$

We form the set R of real numbers as the set of all Dedekind cuts A of Q, and define a total ordering on the real numbers as follows: $x \leq y \Leftrightarrow x \subseteq y$

We embed the rational numbers into the reals by identifying the rational number q with the set of all smaller rational numbers $\{x \in \mathbf{Q} : x < q\}$. Since the rational numbers are dense, such a set can have no greatest element and thus fulfills the conditions for being a real number laid out above.

Addition. $A + B := \{a + b : a \in A \wedge b \in B\}$

Subtraction. $A - B := \{a - b : a \in A \wedge b \in (\mathbf{Q} \setminus B)\}$ where $\mathbf{Q} \setminus B$ denotes the relative complement of B in $\mathbf{Q}$, $\{x : x \in \mathbf{Q} \wedge x \notin B\}$

Negation is a special case of subtraction: $-B := \{a - b : a < 0 \wedge b \in (\mathbf{Q} \setminus B)\}$

Defining multiplication is less straightforward.

if $A, B \geq 0$ then
$A \times B := \{a \times b : a \geq 0 \wedge a \in A \wedge b \geq 0 \wedge b \in B\} \cup \{x \in \mathrm{Q} : x < 0\}$
if either A or B is negative, we use the identities
$A \times B = -(A \times -B) = -(-A \times B) = (-A \times -B)$ to convert
A and/or B to positive numbers and then apply the definition above.

if $A \geq 0$ and $B > 0$ then $A/B := \{a/b : a \in A \wedge b \in (\mathbf{Q} \setminus B)\}$
if either A or B is negative, we use the identities
$A/B = -(A/-B) = -(-A/B) = -A/-B$ to convert A to a non-negative number and/or B to a positive number and then apply the definition above.

Supremum. If a nonempty set S of real numbers has any upper bound in R, then it has a least upper bound in R that is equal to $\bigcup S$.

As an example of a Dedekind cut representing an irrational number, we may take the positive square root of 2. This can be defined by the set

$A = \{x \in \mathbf{Q} : x < 0 \vee x \times x < 2\}$. It can be seen from the definitions above that A is a real number, and that A × A = 2. However, neither claim is immediate. Showing that A is real requires showing that for any positive rational x.

with $x \times x < 2$, there is a rational y with $x < y$ and $y \times y < 2$. The choice $y = \frac{2x+2}{x+2}$ works. Then $A \times A \leq 2$ but to show equality requires showing that if r is any rational number less than 2, then there is positive x in A with $r < x \times x$.

An advantage of this construction is that each real number corresponds to a unique cut.

Stevin's Construction

It has been known since Simon Stevin that real numbers can be represented by decimals. We can take the infinite decimal expansion to be the definition of a real number, **defining** expansions like 0.9999... and 1.0000... to be equivalent, and define the arithmetical operations formally. This is equivalent to the constructions by Cauchy sequences or Dedekind cuts and incorporates an explicit modulus of convergence. Similarly, another radix can be used. Weierstrass attempted to construct the reals but did not entirely succeed. He pointed out that they need only be thought of as complete aggregates (sets) of units and unit fractions.

This construction has the advantage that it is close to the way we are used to thinking of real numbers and suggests series expansions for functions. A standard approach to show that all models of a complete ordered field are isomorphic is to show that any model is isomorphic to this one because we can systematically build a decimal expansion for each element.

Construction Using Hyperreal Numbers

As in the hyperreal numbers, one constructs the hyperrationals $^*\mathbf{Q}$ from the rational numbers by means of an ultrafilter. Here a hyperrational is by definition a ratio of two hyperintegers. Consider the ring B of all limited (i.e. finite) elements in $^*\mathbf{Q}$. Then B has a unique maximal ideal I, the infinitesimal numbers. The quotient ring B/I gives the field **R** of real numbers. Note that B is not an internal set in $^*\mathbf{Q}$. Note that this construction uses a non-principal ultrafilter over the set of natural numbers, the existence of which is guaranteed by the axiom of choice. It turns out that the maximal ideal respects the order on $^*\mathbf{Q}$. Hence the resulting field is an ordered field. Completeness can be proved in a similar way to the construction from the Cauchy sequences.

Construction from Surreal Numbers

Every ordered field can be embedded in the surreal numbers. The real numbers form a maximal subfield that is Archimedean (meaning that no real number is infinitely large). This embedding is not unique, though it can be chosen in a canonical way.

Construction from Z (Eudoxus Reals)

A relatively less known construction allows to define real numbers using only the additive group of integers Z with different versions. The construction has been formally verified by the IsarMathLib project. Shenitzer and Arthan refer to this construction as the Eudoxus reals.

Let an almost homomorphism be a map $f : Z \to Z$ such that the set $\{ f (n + m) - f (m) - f (n) : n , m \in Z \}$ is finite. (Note that $f (n) = \in \ \alpha n \in$ is an almost homomorphism for every $\alpha \in R$. We say that two almost homomorphisms f,g are almost equal if the set $\{ f (n) - g (n) : n \in Z \}$ is finite. This defines an equivalence relation on the set of almost homomorphisms. Real numbers are defined as the equivalence classes of this relation. To add real numbers defined this way we add the almost homomorphisms that represent them. Multiplication of real numbers corresponds to composition of almost homomorphisms. If [f] denotes the real number represented by an almost homomorphism f {\displaystyle f} f we say that $0 \leq [f]$ f is bounded or f takes an infinite number of positive values on $Z +$. This defines the linear order relation on the set of real numbers constructed this way.

Order Properties of the Real Numbers

The real numbers have several important lattice-theoretic properties that are absent in the complex numbers. Most importantly, the real numbers form an ordered field, in which addition and multiplication preserve positivity. Moreover, the ordering of the real numbers is total, and the real numbers have the least upper bound property. These order-theoretic properties lead to a number of important results in real analysis, such as the monotone convergence theorem, the intermediate value theorem and the mean value theorem. However, while the results in real analysis are stated for real numbers, many of these results can be generalized to other mathematical objects. In particular, many ideas in functional analysis and operator theory generalize properties of the real numbers – such generalizations include the theories of Riesz spaces and positive operators. Also, mathematicians consider real and imaginary parts of complex sequences, or by pointwise evaluation of operator sequences.

Sequences

A sequence is usually defined as a function whose domain is a countable totally ordered set, although in many disciplines the domain is restricted, such as to the natural numbers. In real analysis a sequence is a function from a subset of the natural numbers to the real numbers. In other words, a sequence is a map $f(n) : \mathbf{N} \to \mathbf{R}$. We might identify $a_n = f(n)$ for all n or just write $a_n : \mathbf{N} \to \mathbf{R}$.

Limits

A **limit** is the value that a function or sequence "approaches" as the input or index approaches some value. Limits are essential to calculus (and mathematical analysis in general) and are used to define continuity, derivatives, and integrals.

Continuity

A function from the set of real numbers to the real numbers can be represented by a graph in the Cartesian plane; such a function is continuous if, roughly speaking, the graph is a single unbroken curve with no "holes" or "jumps".

Uniform Continuity

If X and Y are subsets of the real numbers, a function $f : X \to Y$ is called **uniformly continuous** if for all $\varepsilon > 0$ there exists a $\delta > 0$ such that for all $x, y \in X$, $|x - y| < \delta$ implies $|f(x) - f(y)| < \varepsilon$.

The difference between being uniformly continuous, and being simply continuous at every point, is that in uniform continuity the value of δ depends only on ε and not on the point in the domain.

Absolute Continuity

Let I be an interval in the real line R. A function $f : I \to R$ is absolutely continuous on I if for every positive number ε, there is a positive number δ such that whenever a finite sequence of pairwise disjoint sub-intervals (x_k, y_k) of I with $x_k, y_k \in I$ satisfies

$$\sum_k (y_k - x_k) < \delta$$

then

$$\sum_k |f(y_k) - f(x_k)| < \varepsilon.$$

The collection of all absolutely continuous functions on I is denoted AC(I).

The following conditions on a real-valued function f on a compact interval [a,b] are equivalent:

(1) f is absolutely continuous;

(2) f has a derivative f' almost everywhere, the derivative is Lebesgue integrable, and

$$f(x) = f(a) + \int_a^x f'(t)\,dt$$

for all x on [a,b];

(3) there exists a Lebesgue integrable function g on [a,b] such that

$$f(x) = f(a) + \int_a^x g(t)\,dt$$

for all x on [a,b].

If these equivalent conditions are satisfied then necessarily $g = f'$ almost everywhere.

Equivalence between (1) and (3) is known as the fundamental theorem of Lebesgue integral calculus, due to Lebesgue.

Series

Given an infinite sequence of numbers { a_n }, a **series** is informally the result of adding all those terms together: $a_1 + a_2 + a_3 + \cdots$. These can be written more compactly using the summation symbol $\sum$. An example is the famous series from Zeno's dichotomy and its mathematical representation:

$$\sum_{n=1}^{\infty} \frac{1}{2^n} = \frac{1}{2} + \frac{1}{4} + \frac{1}{8} + \cdots.$$

The terms of the series are often produced according to a certain rule, such as by a formula, or by an algorithm.

Fourier Series

A **Fourier series** decomposes periodic functions or periodic signals into the sum of a (possibly infinite) set of simple oscillating functions, namely sines and cosines (or complex exponentials). The study of Fourier series is a branch of Fourier analysis.

Differentiation

Formally, the derivative of the function f at a is the limit

$$f'(a) = \lim_{h \to 0} \frac{f(a+h) - f(a)}{h}$$

If the derivative exists everywhere, the function is differentiable. One can take higher derivatives as well, by iterating this process.

One can classify functions by their differentiability class. The class C^0 consists of all continuous functions. The class C^1 consists of all differentiable functions whose derivative is continuous; such functions are called continuously differentiable. Thus, a C^1 function is exactly a function whose derivative exists and is of class C^0. In general, the classes C^k can be defined recursively by declaring C^0 to be the set of all continuous functions and declaring C^k for any positive integer k to be the set of all differentiable functions whose derivative is in C^{k-1}. In particular, C^k is contained in C^{k-1} for every k, and there are examples to show that this containment is strict. C^∞ is the intersection of the sets C^k as k varies over the non-negative integers. C^ω consists of all analytic functions, and is strictly contained in C^∞.

Integration

Riemann Integration

The Riemann integral is defined in terms of Riemann sums of functions with respect to tagged partitions of an interval. Let [a,b] be a closed interval of the real line; then a tagged partition of [a,b] is a finite sequence

$$a = x_0 \le t_1 \le x_1 \le t_2 \le x_2 \le \cdots \le x_{n-1} \le t_n \le x_n = b.$$

This partitions the interval [a, b] into n sub-intervals [xi−1, xi] indexed by i, each of which is "tagged" with a distinguished point ti ∈ [xi−1, xi]. A Riemann sum of a function f with respect to such a tagged partition is defined as

$$\sum_{i=1}^{n} f(t_i)\Delta_i;$$

thus each term of the sum is the area of a rectangle with height equal to the function value at the distinguished point of the given sub-interval, and width the same as the sub-interval width. Let $\Delta_i = x_i - x_{i-1}$ be the width of sub-interval i; then the mesh of such a tagged partition is the width of the largest sub-interval formed by the partition, $\max_{i=1\ldots n} \Delta_i$. The Riemann integral of a function f over the interval [a,b] is equal to S if:

For all ε > 0 there exists δ > 0 such that, for any tagged partition [a,b] with mesh less than δ, we have

$$\left| S - \sum_{i=1}^{n} f(t_i)\Delta_i \right| < \varepsilon.$$

When the chosen tags give the maximum (respectively, minimum) value of each interval, the Riemann sum becomes an upper (respectively, lower) Darboux sum, suggesting the close connection between the Riemann integral and the Darboux integral.

Lebesgue Integration

Lebesgue integration is a mathematical construction that extends the integral to a larger class of functions; it also extends the domains on which these functions can be defined.

Distributions

Distributions (or **generalized functions**) are objects that generalize functions. Distributions make it possible to differentiate functions whose derivatives do not exist in the classical sense. In particular, any locally integrable function has a distributional derivative.

Relation to Complex Analysis

Real analysis is an area of analysis that studies concepts such as sequences and their limits, continuity, differentiation, integration and sequences of functions. By definition, real analysis focuses on the real numbers, often including positive and negative infinity to form the extended real line. Real analysis is closely related to complex analysis, which studies broadly the same properties of complex numbers. In complex analysis, it is natural to define differentiation via holomorphic functions, which have a number of useful properties, such as repeated differentiability, expressability as power series, and satisfying the Cauchy integral formula. In real analysis, it is usually more natural to consider differentiable, smooth, or harmonic functions, which are more widely applicable, but may lack some more powerful properties of holomorphic functions. However, results such as the fundamental theorem of algebra are simpler when expressed in terms of complex numbers.

Techniques from the theory of analytic functions of a complex variable are often used in real analysis – such as evaluation of real integrals by residue calculus.

Important Results

Important results include the Bolzano–Weierstrass and Heine–Borel theorems, the intermediate value theorem and mean value theorem, the fundamental theorem of calculus, and the monotone convergence theorem.

Various ideas from real analysis can be generalized from real space to general metric spaces, as well as to measure spaces, Banach spaces, and Hilbert spaces.

REFERENCES

1. Gaughan, Edward. "1.1 Sequences and Convergence". Introduction to Analysis. AMS (2009). ISBN 0-8218-4787-2.
2. Stewart, James (2008). Calculus: Early Transcendentals (6th ed.). Brooks/Cole. ISBN 0-495-01166-5.
3. Royden 1988, Sect. 5.4, page 108; Nielsen 1997, Definition 15.6 on page 251; Athreya & Lahiri 2006, Definitions 4.4.1, 4.4.2 on pages 128,129. The interval I is assumed to be bounded and closed in the former two books but not the latter book.
4. Nielsen 1997, Theorem 20.8 on page 354; also Royden 1988, Sect. 5.4, page 110 and Athreya & Lahiri 2006, Theorems 4.4.1, 4.4.2 on pages 129,130.
5. Athreya & Lahiri 2006, before Theorem 4.4.1 on page 129.
6. Pugh, Charles Chapman (2002). Real Mathematical Analysis. New York: Springer. pp. 11–15. ISBN 0-387-95297-7.
7. Hersh, Reuben (1997). What is Mathematics, Really?. New York: Oxford University Press US. p. 274. ISBN 0-19-513087-1.
8. Karin Usadi Katz and Mikhail G. Katz (2011) Stevin Numbers and Reality. Foundations of Science. doi:10.1007/s10699-011-9228-9 Online First.
9. V. Dantscher, "Vorlesungen über die Weierstrass'sche Theorie der irrationalen Zahlen" , Teubner (1908)
10. R.D. Arthan. "The Eudoxus Real Numbers". arXiv:math/0405454 .
11. Norbert A'Campo. "A natural construction for the real numbers". arXiv:math/0301015 .
12. Ross Street (September 2003). "Update on the efficient reals" (PDF). Retrieved 2010-10-23.
13. "IsarMathLib".
14. Shenitzer, A. (1987) A topics course in mathematics. The Mathematical

Intelligencer 9, no. 3, 44--52.

15. F. Faltin, N. Metropolis, B. Ross and G.-C. Rota, The real numbers as a wreath product Advances in Math., 16(1975), 278–304.
16. N.G. de Bruijn, N. G. Construction of the system of real numbers. (Dutch) Nederl. Akad. Wetensch. Verslag Afd. Natuurk. 86 (1977), no. 9, 121–125.
17. N. G. de Bruijn,Defining reals without the use of rationals. Nederl. Akad. Wetensch. Proc. Ser. A 79 = Indag. Math. 38 (1976), no. 2, 100–108
18. Rieger, G. J. A new approach to the real numbers (motivated by continued fractions). Abh. Braunschweig.Wiss. Ges. 33 (1982), 205–217
19. Knopfmacher, Arnold; Knopfmacher, John Two concrete new constructions of the real numbers. Rocky Mountain J. Math. 18 (1988), no. 4, 813–824.
20. Knopfmacher, Arnold; Knopfmacher, John A new construction of the real numbers (via infinite products). Nieuw Arch. Wisk. (4) 5 (1987), no. 1, 19–31.
21. MR693180 (84j:26002) review of A new approach to the real numbers (motivated by continued fractions) by Rieger, G. J.

CHAPTER 9

Fourier Analysis

Chapter Contents

- **Introduction**
- **Applications**
- **Fourier Transform**
- **Fourier Series**

Fourier analysis is a method of defining periodic waveform s in terms of trigonometric function s. The method gets its name from a French mathematician and physicist named Jean Baptiste Joseph, Baron de Fourier, who lived during the 18th and 19th centuries. Fourier analysis is used in electronics, acoustics, and communications.

INTRODUCTION

In mathematics, Fourier analysis is the study of the way general functions may be represented or approximated by sums of simpler trigonometric functions. Fourier analysis grew from the study of Fourier series, and is named after Joseph Fourier, who showed that representing a function as a sum of trigonometric functions greatly simplifies the study of heat transfer.

Today, the subject of Fourier analysis encompasses a vast spectrum of mathematics. In the sciences and engineering, the process of decomposing a function into oscillatory components is often called Fourier analysis, while the operation of rebuilding the function from these pieces is known as Fourier synthesis. For example, determining what component frequencies are present in a musical note would involve computing the Fourier transform of a sampled musical note. One could then re-synthesize the same sound by including the frequency components as revealed in the Fourier analysis. In mathematics, the term Fourier analysis often refers to the study of both operations.

The decomposition process itself is called a Fourier transformation. Its output, the Fourier transform, is often given a more specific name, which depends on the domain and other properties of the function being transformed. Moreover, the original concept of Fourier analysis has been extended over time to apply to more and more abstract and general situations, and the general field is often known as harmonic analysis. Each transform used for analysis (see list of Fourier-related transforms) has a corresponding inverse transform that can be used for synthesis.

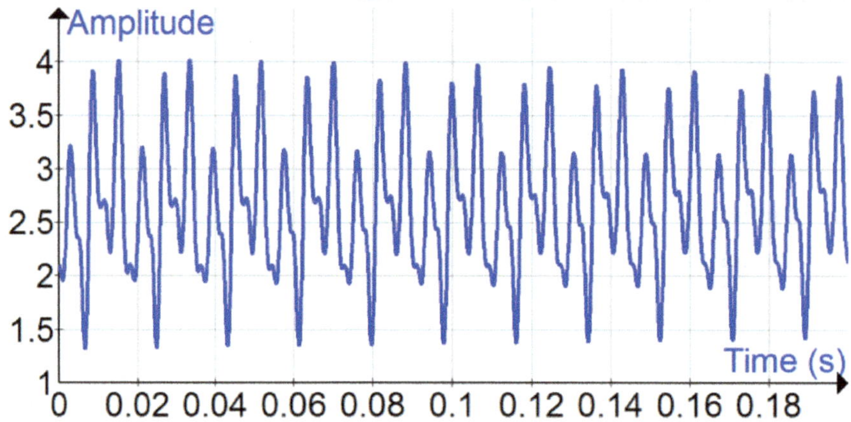

Bass guitar time signal of open string A note (55 Hz).

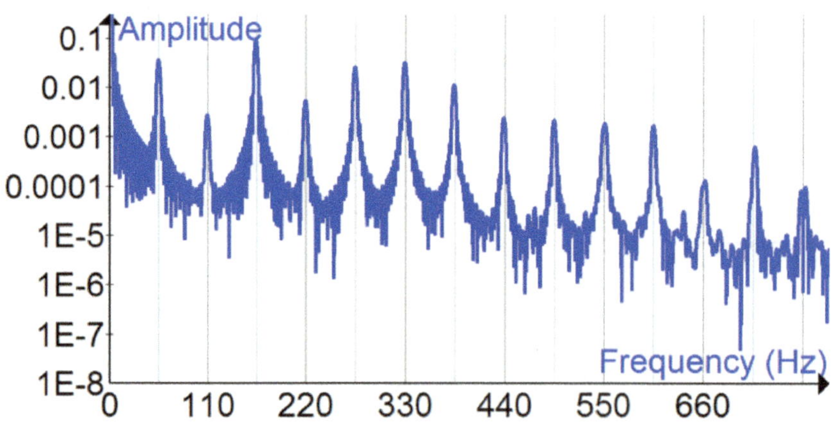

Fourier transform of bass guitar time signal of open string A note (55 Hz), computed with https://sourceforge.net/projects/amoreaccuratefouriertransform/ . Fourier analysis reveals the oscillatory components of signals and functions.

APPLICATIONS

Fourier analysis has many scientific applications – in physics, partial differential equations, number theory, combinatorics, signal processing, imaging, probability theory, statistics, forensics, option pricing, cryptography, numerical analysis, acoustics, oceanography, sonar, optics, diffraction, geometry, protein structure analysis, and other areas.

This wide applicability stems from many useful properties of the transforms:

- The transforms are linear operators and, with proper normalization, are unitary as well (a property known as Parseval's theorem or, more generally, as the Plancherel theorem, and most generally via Pontryagin duality) (Rudin 1990).
- The transforms are usually invertible.
- The exponential functions are eigenfunctions of differentiation, which means that this representation transforms linear differential equations with constant coefficients into ordinary algebraic ones (Evans 1998). Therefore, the behavior of a linear time-invariant system can be analyzed at each frequency independently.
- By the convolution theorem, Fourier transforms turn the complicated convolution operation into simple multiplication, which means that they provide an efficient way to compute convolution-based operations such as polynomial multiplication and multiplying large numbers (Knuth 1997).
- The discrete version of the Fourier transform (see below) can be evaluated quickly on computers using Fast Fourier Transform (FFT) algorithms. (Conte & de Boor 1980)

In forensics, laboratory infrared spectrophotometers use Fourier transform analysis for measuring the wavelengths of light at which a material will absorb in the infrared spectrum. The FT method is used to decode the measured signals and record the wavelength data. And by using a computer, these Fourier calculations are rapidly carried out, so that in a matter of seconds, a computer-operated FT-IR instrument can produce an infrared absorption pattern comparable to that of a prism instrument.

Fourier transformation is also useful as a compact representation of a signal. For example, JPEG compression uses a variant of the Fourier transformation (discrete cosine transform) of small square pieces of a digital image. The Fourier components of each square are rounded to lower arithmetic precision, and weak components are eliminated entirely, so that the remaining components can be stored very compactly. In image reconstruction, each image square is reassembled from the preserved approximate Fourier-transformed components, which are then inverse-transformed to produce an approximation of the original image.

Applications in Signal Processing

When processing signals, such as audio, radio waves, light waves, seismic waves, and even images, Fourier analysis can isolate narrowband components of a compound waveform, concentrating them for easier detection or removal. A large family of signal processing techniques consist of Fourier-transforming a signal, manipulating the Fourier-transformed data in a simple way, and reversing the transformation.

Some examples include:

- Equalization of audio recordings with a series of bandpass filters;
- Digital radio reception without a superheterodyne circuit, as in a modern cell phone or radio scanner;
- Image processing to remove periodic or anisotropic artifacts such as jaggies from interlaced video, stripe artifacts from strip aerial photography, or wave patterns from radio frequency interference in a digital camera;
- Cross correlation of similar images for co-alignment;
- X-ray crystallography to reconstruct a crystal structure from its diffraction pattern;
- Fourier transform ion cyclotron resonance mass spectrometry to determine the mass of ions from the frequency of cyclotron motion in a magnetic field;
- Many other forms of spectroscopy, including infrared and nuclear magnetic resonance spectroscopies;
- Generation of sound spectrograms used to analyze sounds;
- Passive sonar used to classify targets based on machinery noise.

Variants of Fourier Analysis

(Continuous) Fourier transform

Most often, the unqualified term Fourier transform refers to the transform of functions of a continuous real argument, and it produces a continuous function of frequency, known as a frequency distribution. One function is transformed into another, and the operation is reversible. When the domain of the input (initial) function is time (t), and the domain of the output (final) function is ordinary frequency, the transform of function s(t) at frequency f is given by the complex number:

$$S(f) = \int_{-\infty}^{\infty} s(t) \cdot e^{-2i\pi ft}\, dt.$$

Evaluating this quantity for all values of f produces the frequency-domain function. Then s(t) can be represented as a recombination of complex exponentials of all possible frequencies:

$$s(t) = \int_{-\infty}^{\infty} S(f) \cdot e^{2i\pi ft}\, df,$$

which is the inverse transform formula. The complex number, S(f), conveys both amplitude and phase of frequency f.

See Fourier transform for much more information, including:

- conventions for amplitude normalization and frequency scaling/units
- transform properties

- tabulated transforms of specific functions
- an extension/generalization for functions of multiple dimensions, such as images.

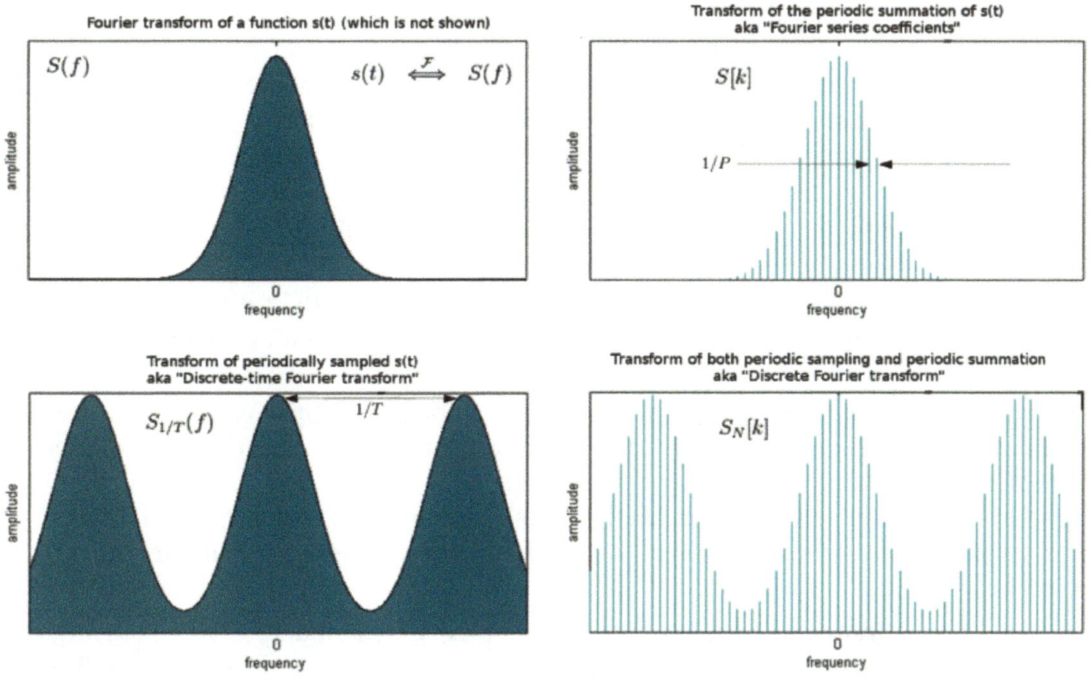

A Fourier transform and 3 variations caused by periodic sampling (at interval T) and/or periodic summation (at interval P) of the underlying time-domain function. The relative computational ease of the DFT sequence and the insight it gives into S(f) make it a popular analysis tool.

FOURIER TRANSFORM

The **Fourier transform** decomposes a function of time (a signal) into the frequencies that make it up, in a way similar to how a musical chord can be expressed as the amplitude (or loudness) of its constituent notes. The Fourier transform of a function of time itself is a complex-valued function of frequency, whose absolute value represents the amount of that frequency present in the original function, and whose complex argument is the phase offset of the basic sinusoid in that frequency. The Fourier transform is called the frequency domain representation of the original signal. The term Fourier transform refers to both the frequency domain representation and the mathematical operation that associates the frequency domain representation to a function of time. The Fourier transform is not limited to functions of time, but in order to have a unified language, the domain of the original function is commonly referred to as the time domain. For many functions of practical interest, one can define an operation that reverses this: the inverse Fourier transformation, also called Fourier synthesis, of a frequency domain representation combines the contributions of all the different frequencies to recover the original function of time.

Linear operations performed in one domain (time or frequency) have corresponding operations in the other domain, which are sometimes easier to perform. The operation of differentiation in the time domain corresponds to multiplication by the frequency, so some differential equations are easier to analyze in the frequency domain. Also, convolution in the time domain corresponds to ordinary multiplication in the frequency domain. Concretely, this means that any linear time-invariant system, such as a filter applied to a signal, can be expressed relatively simply as an operation on frequencies. After performing the desired operations, transformation of the result can be made back to the time domain. Harmonic analysis is the systematic study of the relationship between the frequency and time domains, including the kinds of functions or operations that are "simpler" in one or the other, and has deep connections to almost all areas of modern mathematics.

Functions that are localized in the time domain have Fourier transforms that are spread out across the frequency domain and vice versa, a phenomenon known as the uncertainty principle. The critical case for this principle is the Gaussian function, of substantial importance in probability theory and statistics as well as in the study of physical phenomena exhibiting normal distribution (e.g., diffusion). The Fourier transform of a Gaussian function is another Gaussian function. Joseph Fourier introduced the transform in his study of heat transfer, where Gaussian functions appear as solutions of the heat equation.

The Fourier transform can be formally defined as an improper Riemann integral, making it an integral transform, although this definition is not suitable for many applications requiring a more sophisticated integration theory. For example, many relatively simple applications use the Dirac delta function, which can be treated formally as if it were a function, but the justification requires a mathematically more sophisticated viewpoint. The Fourier transform can also be generalized to functions of several variables on Euclidean space, sending a function of 3-dimensional space to a function of 3-dimensional momentum (or a function of space and time to a function of 4-momentum). This idea makes the spatial Fourier transform very natural in the study of waves, as well as in quantum mechanics, where it is important to be able to represent wave solutions as functions of either space or momentum and sometimes both. In general, functions to which Fourier methods are applicable are complex-valued, and possibly vector-valued. Still further generalization is possible to functions on groups, which, besides the original Fourier transform on $\mathbb{R}$ or $\mathbb{R}^n$ (viewed as groups under addition), notably includes the discrete-time Fourier transform (DTFT, group = $\mathbb{Z}$), the discrete Fourier transform (DFT, group = $\mathbb{Z}$ mod N) and the Fourier series or circular Fourier transform (group = S^1, the unit circle ≈ closed finite interval with endpoints identified). The latter is routinely employed to handle periodic functions. The fast Fourier transform (FFT) is an algorithm for computing the DFT.

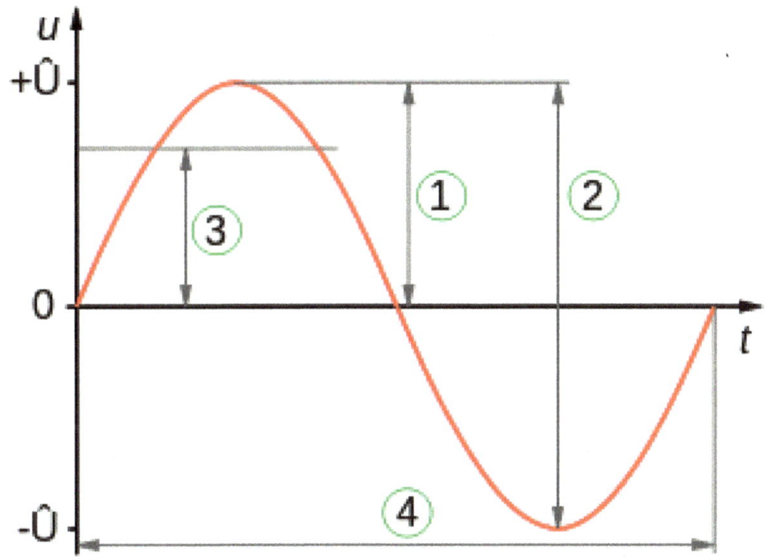

A sinusoidal curve, with peak amplitude (1).

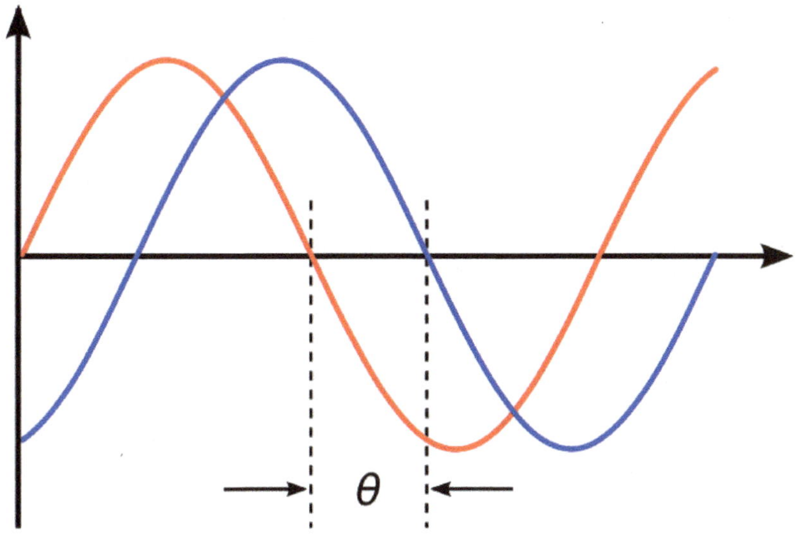

Illustration of phase shift θ.

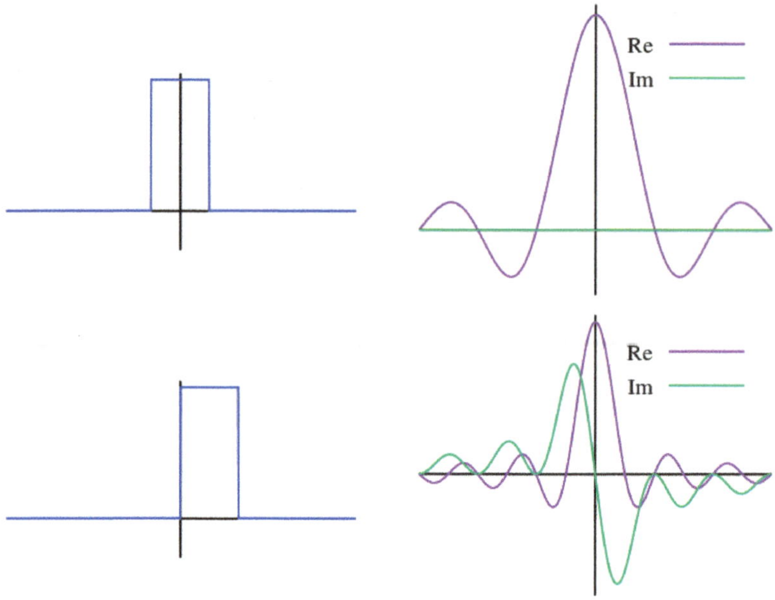

In the first row is the graph of the unit pulse function f(t) and its fourier tansform f(ω), a function of frequency ω Translation (that is, delay) in the time domain goes over to complex phase shifts in the frequency domain. In the second row is shown g(t) a delayed unit pulse, beside the real and imaginary parts of the Fourier transform. The Fourier transform decomposes a function into eigenfunctions for the group of translations.

Definition

The Fourier transform of the function f is traditionally denoted by adding a circumflex: f. There are several common conventions for defining the Fourier transform of an integrable function f : $\mathbb{R} \to$ C.[1][2] Here we will use the following definition:

$$\hat{f}(\xi) = \int_{-\infty}^{\infty} f(x)\, e^{-2\pi i x \xi}\, dx,$$

for any real number ξ.

When the independent variable x represents time (with SI unit of seconds), the transform variable ξ represents frequency (in hertz). Under suitable conditions, f is determined by f via the inverse transform:

$$f(x) = \int_{-\infty}^{\infty} \hat{f}(\xi)\, e^{2\pi i \xi x}\, d\xi,$$

for any real number x.

The statement that f can be reconstructed from f is known as the Fourier inversion theorem, and was first introduced in Fourier's Analytical Theory of Heat, although what would be considered a proof by modern standards was not given until much later. The functions f and f often are referred to as a Fourier integral pair or Fourier transform pair.

For other common conventions and notations, including using the angular frequency ω instead of the frequency ξ, see Other conventions and Other notations below. The

Fourier transform on Euclidean space is treated separately, in which the variable x often represents position and ξ momentum. The conventions chosen in this article are those of harmonic analysis, and are characterized as the unique conventions such that the Fourier transform is both unitary on $L^2$and an algebra homomorphism from L^1 to L^∞ without normalizing the Lebesgue measure.

Many other characterizations of the Fourier transform exist. For example, one uses the Stone–von Neumann theorem: the Fourier transform is the unique unitary intertwiner for the symplectic and Euclidean Schrödinger representations of the Heisenberg group.

One motivation for the Fourier transform comes from the study of Fourier series. In the study of Fourier series, complicated but periodic functions are written as the sum of simple waves mathematically represented by sines and cosines. The Fourier transform is an extension of the Fourier series that results when the period of the represented function is lengthened and allowed to approach infinity.

Due to the properties of sine and cosine, it is possible to recover the amplitude of each wave in a Fourier series using an integral. In many cases it is desirable to use Euler's formula, which states that $e^{2\pi i\theta} = \cos(2\pi\theta) + i \sin(2\pi\theta)$, to write Fourier series in terms of the basic waves $e^{2\pi i\theta}$. This has the advantage of simplifying many of the formulas involved, and provides a formulation for Fourier series that more closely resembles the definition followed in this article. Re-writing sines and cosines as complex exponentials makes it necessary for the Fourier coefficients to be complex valued. The usual interpretation of this complex number is that it gives both the amplitude (or size) of the wave present in the function and the phase (or the initial angle) of the wave. These complex exponentials sometimes contain negative "frequencies". If θ is measured in seconds, then the waves $e^{2\pi i\theta}$ and $e^{-2\pi i\theta}$ both complete one cycle per second, but they represent different frequencies in the Fourier transform. Hence, frequency no longer measures the number of cycles per unit time, but is still closely related.

There is a close connection between the definition of Fourier series and the Fourier transform for functions f that are zero outside an interval. For such a function, we can calculate its Fourier series on any interval that includes the points where f is not identically zero. The Fourier transform is also defined for such a function. As we increase the length of the interval on which we calculate the Fourier series, then the Fourier series coefficients begin to look like the Fourier transform and the sum of the Fourier series of f begins to look like the inverse Fourier transform. To explain this more precisely, suppose that T is large enough so that the interval [−T/2, T/2] contains the interval on which f is not identically zero. Then the n-th series coefficient c_n is given by:

$$c_n = \frac{1}{T}\int_{-\frac{T}{2}}^{\frac{T}{2}} f(x)\, e^{-2\pi i\left(\frac{n}{T}\right)x}\, dx.$$

Comparing this to the definition of the Fourier transform, it follows that

$$c_n = \frac{1}{T}\hat{f}\left(\frac{n}{T}\right)$$

since $f(x)$ is zero outside $[-\frac{T}{2}, \frac{T}{2}]$. Thus the Fourier coefficients are just the values of the Fourier transform sampled on a grid of width $\frac{1}{T}$, multiplied by the grid width $\frac{1}{T}$.

Under appropriate conditions, the Fourier series of f will equal the function f. In other words, f can be written:

$$f(x) = \sum_{n=-\infty}^{\infty} c_n \, e^{2\pi i \left(\frac{n}{T}\right)x} = \sum_{n=-\infty}^{\infty} \hat{f}\left(\xi_n\right) e^{2\pi i \xi_n x} \, \Delta\xi,$$

where the last sum is simply the first sum rewritten using the definitions $\xi_n = n/T$, and $\Delta\xi = (n + 1)/T - n/T = 1/T$.

This second sum is a Riemann sum, and so by letting $T \to \infty$ it will converge to the integral for the inverse Fourier transform given in the definition section. Under suitable conditions, this argument may be made precise.

In the study of Fourier series the numbers c_n could be thought of as the "amount" of the wave present in the Fourier series of f. Similarly, as seen above, the Fourier transform can be thought of as a function that measures how much of each individual frequency is present in our function f, and we can recombine these waves by using an integral (or "continuous sum") to reproduce the original function. $\hat{f}(3)$ we must integrate $e^{-2\pi i(3t)}f(t)$. The second image shows the plot of the real and imaginary parts of this function. The real part of the integrand is almost always positive, because when f(t) is negative, the real part of $e^{-2\pi i(3t)}$ is negative as well. Because they oscillate at the same rate, when f(t) is positive, so is the real part of $e^{-2\pi i(3t)}$. The result is that when you integrate the real part of the integrand you get a relatively large number (in this case 0.5). On the other hand, when you try to measure a frequency that is not present, as in the case when we look at $\hat{f}(5)$, you see that both real and imaginary component of this function vary rapidly between positive and negative values, as plotted in the third image. Therefore, in this case, the integrand oscillates fast enough so that the integral is very small and the value for the Fourier transform for that frequency is nearly zero. The general situation may be a bit more complicated than this, but this in spirit is how the Fourier transform measures how much of an individual frequency is present in a function f(t).

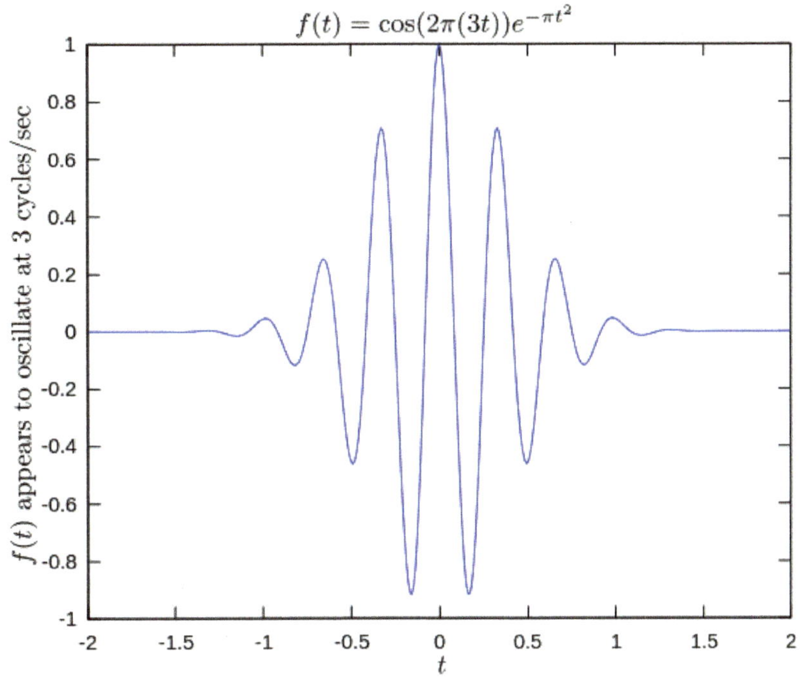

Original function showing oscillation 3 Hz.

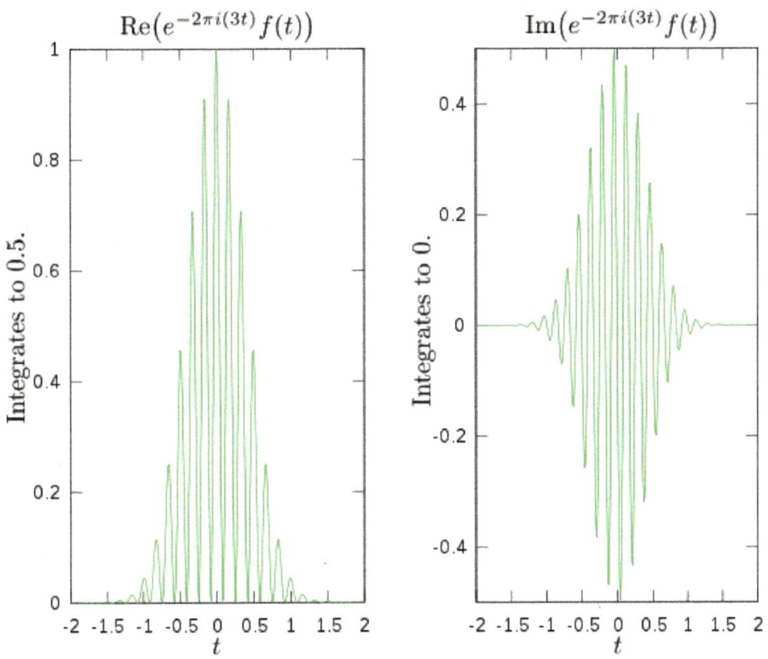

Real and imaginary parts of integrand for Fourier transform at 3 Hz

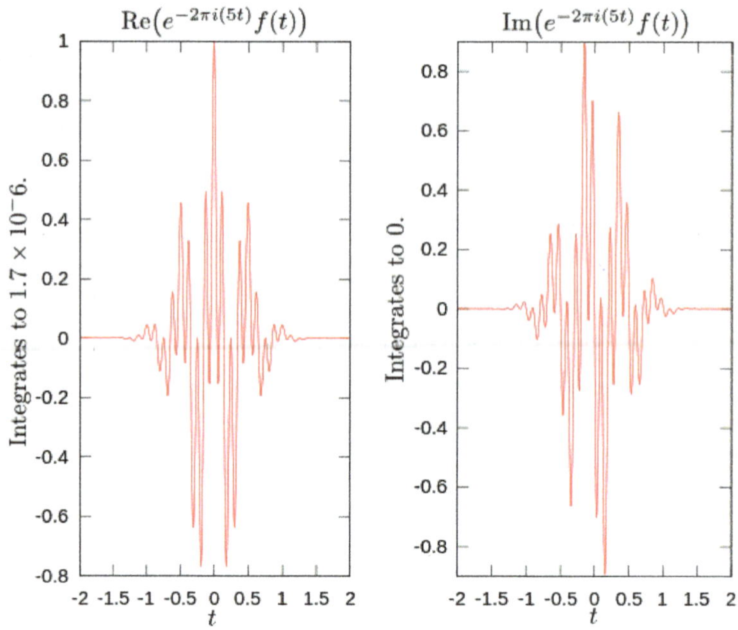

Real and imaginary parts of integrand for Fourier transform at 5 Hz

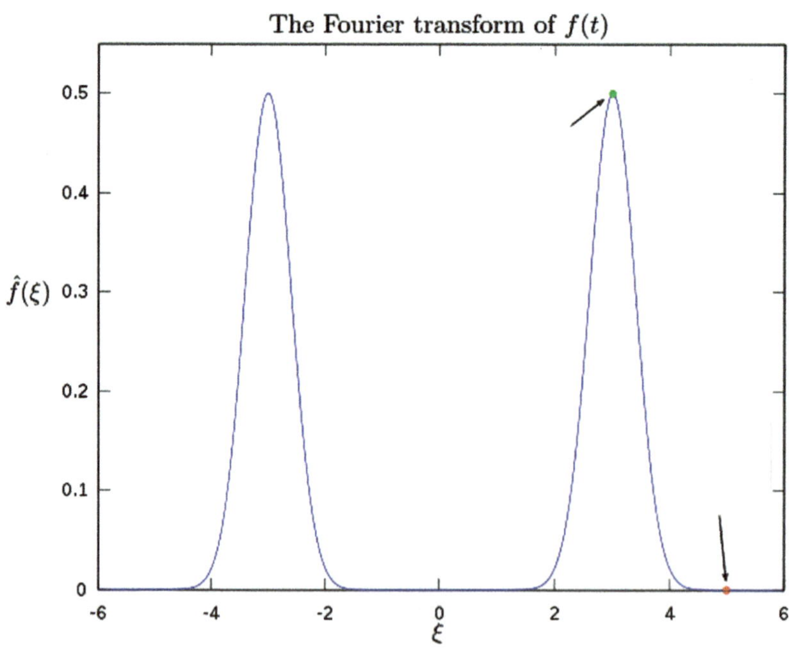

Fourier transform with 3 and 5 Hz labeled

Invertibility and Periodicity

Under suitable conditions on the function f, it can be recovered from its Fourier transform f Indeed, denoting the Fourier transform operator by F, so F(f) := f then for suitable functions, applying the Fourier transform twice simply flips the function $\mathcal{F}^2(f)(x) = f(-x)$, which can be interpreted as "reversing time". Since reversing time is two-periodic, applying this twice yields F^4 (f) = f, so the Fourier transform operator is four-periodic, and similarly the inverse Fourier transform can be obtained by applying the Fourier transform three times: F^3 (f) = f. In particular the Fourier transform is invertible (under suitable conditions).

More precisely, defining the parity operator P that inverts time, $\mathcal{P}[f] : t \mapsto f(-t)$:

$$\mathcal{F}^0 = \mathrm{Id}, \qquad \mathcal{F}^1 = \mathcal{F}, \qquad \mathcal{F}^2 = \mathcal{P}, \qquad \mathcal{F}^4 = \mathrm{Id}$$

$$\mathcal{F}^3 = \mathcal{F}^{-1} = \mathcal{P} \circ \mathcal{F} = \mathcal{F} \circ \mathcal{P}$$

These equalities of operators require careful definition of the space of functions in question, defining equality of functions (equality at every point? equality almost everywhere?) and defining equality of operators – that is, defining the topology on the function space and operator space in question. These are not true for all functions, but are true under various conditions, which are the content of the various forms of the Fourier inversion theorem.

This four-fold periodicity of the Fourier transform is similar to a rotation of the plane by 90°, particularly as the two-fold iteration yields a reversal, and in fact this analogy can be made precise. While the Fourier transform can simply be interpreted as switching the time domain and the frequency domain, with the inverse Fourier transform switching them back, more geometrically it can be interpreted as a rotation by 90° in the time–frequency domain (considering time as the x-axis and frequency as the y-axis), and the Fourier transform can be generalized to the fractional Fourier transform, which involves rotations by other angles. This can be further generalized to linear canonical transformations, which can be visualized as the action of the special linear group $SL_2(\mathbf{R})$ on the time–frequency plane, with the preserved symplectic form corresponding to the uncertainty principle, below. This approach is particularly studied in signal processing, under time–frequency analysis.

Units and Duality

In mathematics, one often does not think of any units as being attached to the two variables t and ξ. But in physical applications, ξ must have inverse units to the units of t. For example, if t is measured in seconds, ξ should be in cycles per second for the formulas here to be valid. If the scale of t is changed and t is measured in units of 2π seconds, then either ξ must be in the so-called "angular frequency", or one must insert some constant scale factor into some of the formulas. If t is measured in units of length, then ξ must be in inverse length, e.g., wavenumbers. That is to say, there are two copies of the real line: one measured in one set of units, where t ranges, and the other in inverse units to the units of t, and which is the range of ξ. So these are two distinct copies of the real line, and cannot be identified with each other. Therefore, the Fourier transform goes from one space of functions to a different space of functions: functions which have a different domain of definition.

In general, ξ must always be taken to be a linear form on the space of ts, which is to say that the second real line is the dual space of the first real line. See the article on linear algebra for a more formal explanation and for more details. This point of view becomes essential in generalisations of the Fourier transform to general symmetry groups, including the case of Fourier series.

That there is no one preferred way (often, one says "no canonical way") to compare the two copies of the real line which are involved in the Fourier transform—fixing the units on one line does not force the scale of the units on the other line—is the reason for the plethora of rival conventions on the definition of the Fourier transform. The various definitions resulting from different choices of units differ by various constants. If the units of t are in seconds but the units of ξ are in angular frequency, then the angular frequency variable is often denoted by one or another Greek letter, for example, $\omega = 2\pi\xi$ is quite common. Thus (writing $\hat{x}_1$ for the alternative definition and $\hat{x}$ for the definition adopted in this article)

$$\hat{x}_1(\omega) = \hat{x}\left(\frac{\omega}{2\pi}\right) = \int_{-\infty}^{\infty} x(t)e^{-i\omega t}\,dt$$

as before, but the corresponding alternative inversion formula would then have to be

$$x(t) = \frac{1}{2\pi}\int_{-\infty}^{\infty} \hat{x}_1(\omega)e^{it\omega}\,d\omega.$$

To have something involving angular frequency but with greater symmetry between the Fourier transform and the inversion formula, one very often sees still another alternative definition of the Fourier transform, with a factor of $\sqrt{2\pi}$, thus

$$\hat{x}_2(\omega) = \frac{1}{\sqrt{2\pi}}\int_{-\infty}^{\infty} x(t)e^{-i\omega t}\,dt,$$

and the corresponding inversion formula then has to be

$$x(t) = \frac{1}{\sqrt{2\pi}}\int_{-\infty}^{\infty} \hat{x}_2(\omega)e^{it\omega}\,d\omega.$$

Furthermore, there is no way to fix which square root of negative one will be meant by the symbol i (it makes no sense to speak of "the positive square root" since only real numbers can be positive, similarly it makes no sense to say "rotation counter-clockwise", because until i is chosen, there is no fixed way to draw the complex plane), and hence one occasionally sees the Fourier transform written with i in the exponent instead of −i, and vice versa for the inversion formula, a convention that is equally valid as the one chosen in this article, which is the more usual one.

For example, in probability theory, the characteristic function ϕ of the probability density function f of a random variable X of continuous type is defined without a negative sign in the exponential, and since the units of x are ignored, there is no 2π either:

$$\phi(\lambda) = \int_{-\infty}^{\infty} f(x) e^{i\lambda x}\, dx.$$

(In probability theory, and in mathematical statistics, the use of the Fourier—Stieltjes transform is preferred, because so many random variables are not of continuous type, and do not possess a density function, and one must treat discontinuous distribution functions, i.e., measures which possess "atoms".)

From the higher point of view of group characters, which is much more abstract, all these arbitrary choices disappear, as will be explained in the later section of this article, on the notion of the Fourier transform of a function on an Abelian locally compact group.

Computation Methods

The appropriate computation method largely depends how the original mathematical function is represented and the desired form of the output function.

Since the fundamental definition of a Fourier transform is an integral, functions that can be expressed as closed-form expressions are commonly computed by working the integral analytically to yield a closed-form expression in the Fourier transform conjugate variable as the result. This is the method used to generate tables of Fourier transforms, including those found in the table below (Fourier transform#Tables of important Fourier transforms).

Many computer algebra systems such as Matlab and Mathematica that are capable of symbolic integration are capable of computing Fourier transforms analytically. For example, to compute the Fourier transform of f(t) = cos(6πt) $e^{-\pi t^2}$ one might enter the command "integrate cos(6*pi*t) exp(−pi*t^2) exp(-i*2*pi*f*t) from -inf to inf» into Wolfram Alpha.

Numerical Integration of Closed-Form Functions

If the input function is in closed-form and the desired output function is a series of ordered pairs (for example a table of values from which a graph can be generated) over a specified domain, then the Fourier transform can be generated by numerical integration at each value of the Fourier conjugate variable (frequency, for example) for which a value of the output variable is desired. Note that this method requires computing a separate numerical integration for each value of frequency for which a value of the Fourier transform is desired. The numerical integration approach works on a much broader class of functions than the analytic approach, because it yields results for functions that do not have closed form Fourier transform integrals.

Numerical Integration of a Series of Ordered Pairs

If the input function is a series of ordered pairs (for example, a time series from measuring an output variable repeatedly over a time interval) then the output function must also be a series of ordered pairs (for example, a complex number vs. frequency over a specified domain of frequencies), unless certain assumptions and approximations are made allowing the output function to be approximated by a closed-form expression. In the general case where the available input series of ordered pairs are assumed be samples representing a continuous function over an interval (amplitude vs. time, for example), the series of ordered pairs representing the desired output function can be obtained by numerical integration of the input data over the available interval at each value of the Fourier conjugate variable (frequency, for example) for which the value of the Fourier transform is desired.

Explicit numerical integration over the ordered pairs can yield the Fourier transform output value for any desired value of the conjugate Fourier transform variable (frequency, for example), so that a spectrum can be produced at any desired step size and over any desired variable range for accurate determination of amplitudes, frequencies, and phases corresponding to isolated peaks. Unlike limitations in DFT and FFT methods, explicit numerical integration can have any desired step size and compute the Fourier transform over any desired range of the congugate Fourier transform variable (for example, frequency).

Discrete Fourier Transforms and Fast Fourier Transforms

If the ordered pairs representing the original input function are equally spaced in their input variable (for example, equal time steps), then the Fourier transform is known as a discrete Fourier transform (DFT), which can be computed either by explicit numerical integration, by explicit evaluation of the DFT definition, or by fast Fourier transform (FFT) methods. In contrast to explicit integration of input data, use of the DFT and FFT methods produces Fourier transforms described by ordered pairs of step size equal to the reciprocal of the original sampling interval. For example, if the input data is sampled for 10 seconds, the output of DFT and FFT methods will have a 0.1 Hz frequency spacing.

FOURIER SERIES

The Fourier transform of a periodic function, $s_P(t)$, with period P, becomes a Dirac comb function, modulated by a sequence of complex coefficients:

$$S[k] = \frac{1}{P}\int_P s_P(t)\cdot e^{-2i\pi\frac{k}{P}t}\,dt$$

for all integer values of k, and where ∫P is the integral over any interval of length P.

The inverse transform, known as **Fourier series**, is a representation of $s_P(t)$ in terms of a summation of a potentially infinite number of harmonically related sinusoids or complex exponential functions, each with an amplitude and phase specified by one of the coefficients:

$$s_P(t) = \sum_{k=-\infty}^{\infty} S[k]\cdot e^{2i\pi\frac{k}{P}t} \quad \overset{\mathcal{F}}{\Longleftrightarrow} \quad \sum_{k=-\infty}^{+\infty} S[k]\,\delta\left(f-\frac{k}{P}\right).$$

When sP(t), is expressed as a periodic summation of another function, s(t):

$$s_P(t) \stackrel{\text{def}}{=} \sum_{k=-\infty}^{\infty} s(t - kP),$$

the coefficients are proportional to samples of S(f) at discrete intervals of 1/P

$$S[k] = \frac{1}{P} \cdot S\left(\frac{k}{P}\right).$$

A sufficient condition for recovering s(t) (and therefore S(f)) from just these samples is that the non-zero portion of s(t) be confined to a known interval of duration P, which is the frequency domain dual of the Nyquist–Shannon sampling theorem.

See Fourier series for more information, including the historical development.

Discrete-Time Fourier Transform (Dtft)

The DTFT is the mathematical dual of the time-domain Fourier series. Thus, a convergent periodic summation in the frequency domain can be represented by a Fourier series, whose coefficients are samples of a related continuous time function:

$$S_{\frac{1}{T}}(f) \stackrel{\text{def}}{=} \underbrace{\sum_{k=-\infty}^{\infty} S\left(f - \frac{k}{T}\right) \equiv \overbrace{\sum_{n=-\infty}^{\infty} s[n] \cdot e^{-2i\pi fnT}}^{\text{Fourier series (DTFT)}}}_{\text{Poisson summation formula}} = \mathcal{F}\left\{\sum_{n=-\infty}^{\infty} s[n]\, \delta(t - nT)\right\}$$

which is known as the DTFT. Thus the DTFT of the s[n] sequence is also the Fourier transform of the modulated Dirac comb function.[note 2]

The Fourier series coefficients (and inverse transform), are defined by:

$$s[n] \stackrel{\text{def}}{=} T \int_{\frac{1}{T}} S_{\frac{1}{T}}(f) \cdot e^{2i\pi fnT}\, df = T \underbrace{\int_{-\infty}^{\infty} S(f) \cdot e^{2i\pi fnT}\, df}_{\stackrel{\text{def}}{=}\, s(nT)}$$

Parameter T corresponds to the sampling interval, and this Fourier series can now be recognized as a form of the Poisson summation formula. Thus we have the important result that when a discrete data sequence, s[n], is proportional to samples of an underlying continuous function, s(t), one can observe a periodic summation of the continuous Fourier transform, S(f). That is a cornerstone in the foundation of digital signal processing. Furthermore, under certain idealized conditions one can theoretically recover S(f) and s(t) exactly. A sufficient condition for perfect recovery is that the non-zero portion of S(f) be confined to a known frequency interval of width 1/T. When that interval is [-0.5/T, 0.5/T], the applicable reconstruction formula is the Whittaker–Shannon interpolation formula.

Another reason to be interested in $S_{1/T}(f)$ is that it often provides insight into the amount of aliasing caused by the sampling process.

Applications of the DTFT are not limited to sampled functions. See Discrete-time Fourier transform for more information on this and other topics, including:

- normalized frequency units
- windowing (finite-length sequences)
- transform properties
- tabulated transforms of specific functions

Discrete Fourier Transform (DFT)

Similar to a Fourier series, the DTFT of a periodic sequence, $s_N[n]$, with period **N**, becomes a Dirac comb function, modulated by a sequence of complex coefficients (see DTFT/Periodic data):

$$S[k] = \sum_N s_N[n] \cdot e^{-2i\pi \frac{k}{N} n},$$

where ∑N is the sum over any n-sequence of length N.

The S[k] sequence is what is customarily known as the DFT of sN. It is also N-periodic, so it is never necessary to compute more than N coefficients. The inverse transform is given by:

$$s_N[n] = \frac{1}{N} \sum_N S[k] \cdot e^{2i\pi \frac{n}{N} k},$$

where ∑N is the sum over any k-sequence of length N.

When sN[n] is expressed as a periodic summation of another function:

$$s_N[n] \stackrel{\text{def}}{=} \sum_{k=-\infty}^{\infty} s[n-kN], \quad \text{and} \quad s[n] \stackrel{\text{def}}{=} s(nT),$$

the coefficients are proportional to samples of $S_{\frac{1}{T}}(f)$ at discrete intervals of $\frac{1}{P} = \frac{1}{NT}$:

$$S[k] = \frac{1}{T} \cdot S_{\frac{1}{T}}\left(\frac{k}{P}\right).$$

Conversely, when one wants to compute an arbitrary number (N) of discrete samples of one cycle of a continuous DTFT, $S_{1/T}(f)$, it can be done by computing the relatively simple DFT of $s_N[n]$, as defined above. In most cases, N is chosen equal to the length of non-zero portion of s[n]. Increasing N, known as zero-padding or interpolation, results in more closely spaced samples of one cycle of $S_{1/T}(f)$. Decreasing N, causes overlap (adding) in the time-domain (analogous to aliasing), which corresponds to decimation in the frequency domain. (see Sampling the DTFT) In most cases of practical interest, the s[n] sequence represents a longer sequence that was truncated by the application of a finite-length window function or FIR filter array.

The DFT can be computed using a fast Fourier transform (FFT) algorithm, which makes it a practical and important transformation on computers.

See Discrete Fourier transform for much more information, including:

- transform properties
- applications
- tabulated transforms of specific functions

Summary

For periodic functions, both the Fourier transform and the DTFT comprise only a discrete set of frequency components (Fourier series), and the transforms diverge at those frequencies. One common practice (not discussed above) is to handle that divergence via Dirac delta and Dirac comb functions. But the same spectral information can be discerned from just one cycle of the periodic function, since all the other cycles are identical. Similarly, finite-duration functions can be represented as a Fourier series, with no actual loss of information except that the periodicity of the inverse transform is a mere artifact. We also note that none of the formulas here require the duration of s to be limited to the period, **P** or **N**. But that is a common situation, in practice.

$s(t)$ transforms (continuous-time)

	Continuous frequency	Discrete frequencies
Transform	$S(f) \stackrel{\text{def}}{=} \int_{-\infty}^{\infty} s(t)\cdot e^{-2i\pi ft}\,dt$	$\overbrace{\frac{1}{P}\cdot S\left(\frac{k}{P}\right)}^{S[k]} \stackrel{\text{def}}{=} \frac{1}{P}\int_{-\infty}^{\infty} s(t)\cdot e^{-2i\pi \frac{k}{P}t}\,dt \equiv \frac{1}{P}\int_P s_P(t)\cdot e^{-2i\pi\frac{k}{P}t}\,dt$
Inverse	$s(t)=\int_{-\infty}^{\infty} S(f)\cdot e^{2i\pi ft}\,df$	$\underbrace{s_P(t)=\sum_{k=-\infty}^{\infty} S[k]\cdot e^{2i\pi\frac{k}{P}t}}_{\text{Poisson summation formula (Fourier series)}}$

$s(nT)$ transforms (discrete-time)

	Continuous frequency	Discrete frequencies
Transform	$\underbrace{\frac{1}{T}S_{\frac{1}{T}}(f) \stackrel{\text{def}}{=} \sum_{n=-\infty}^{\infty} s(nT)\cdot e^{-2i\pi fnT}}_{\text{Poisson summation formula (DTFT)}}$	$\overbrace{\frac{1}{T}S_{\frac{1}{T}}\left(\frac{k}{NT}\right)}^{S[k]} \stackrel{\text{def}}{=} \sum_{n=-\infty}^{\infty} s(nT)\cdot e^{-2i\pi\frac{kn}{N}}$ $\equiv \underbrace{\sum_N s_P(nT)\cdot e^{-2i\pi\frac{kn}{N}}}_{\text{DFT}}$
Inverse	$s(nT)=T\int_{\frac{1}{T}} \frac{1}{T}S_{\frac{1}{T}}(f)\cdot e^{2i\pi fnT}\,df$ $\sum_{n=-\infty}^{\infty} s(nT)\cdot\delta(t-nT) = \underbrace{\int_{-\infty}^{\infty}\frac{1}{T}S_{\frac{1}{T}}(f)\cdot e^{2i\pi ft}\,df}_{\text{inverse Fourier transform}}$	$s_P(nT)=\overbrace{\frac{1}{N}\sum_N S[k]\cdot e^{2i\pi\frac{kn}{N}}}^{\text{inverse DFT}}$ $=\frac{1}{P}\sum_N S_{\frac{1}{T}}\left(\frac{k}{P}\right)\cdot e^{2i\pi\frac{kn}{N}}$

Fourier Transforms on Arbitrary Locally Compact Abelian Topological Groups

The Fourier variants can also be generalized to Fourier transforms on arbitrary locally compact abelian topological groups, which are studied in harmonic analysis; there, the Fourier transform takes functions on a group to functions on the dual group. This treatment also allows a general formulation of the convolution theorem, which relates Fourier transforms and convolutions. The Pontryagin duality for the generalized underpinnings of the Fourier transform.

Time–Frequency Transforms

In signal processing terms, a function (of time) is a representation of a signal with perfect time resolution, but no frequency information, while the Fourier transform has perfect frequency resolution, but no time information.

As alternatives to the Fourier transform, in time–frequency analysis, one uses time–frequency transforms to represent signals in a form that has some time information and some frequency information – by the uncertainty principle, there is a trade-off between these. These can be generalizations of the Fourier transform, such as the short-time Fourier transform, the Gabor transform or fractional Fourier transform (FRFT), or can use different functions to represent signals, as in wavelet transforms and chirplet transforms, with the wavelet analog of the (continuous) Fourier transform being the continuous wavelet transform.

A primitive form of harmonic series dates back to ancient Babylonian mathematics, where they were used to compute ephemerides (tables of astronomical positions).

The classical Greek concepts of deferent and epicycle in the Ptolemaic system of astronomy were related to Fourier series (see Deferent and epicycle: Mathematical formalism).

In modern times, variants of the discrete Fourier transform were used by Alexis Clairaut in 1754 to compute an orbit, which has been described as the first formula for the DFT, and in 1759 by Joseph Louis Lagrange, in computing the coefficients of a trigonometric series for a vibrating string. Technically, Clairaut's work was a cosine-only series (a form of discrete cosine transform), while Lagrange's work was a sine-only series (a form of discrete sine transform); a true cosine+sine DFT was used by Gauss in 1805 for trigonometric interpolation of asteroid orbits. Euler and Lagrange both discretized the vibrating string problem, using what would today be called samples.

An early modern development toward Fourier analysis was the 1770 paper Réflexions sur la résolution algébrique des équations by Lagrange, which in the method of Lagrange resolvents used a complex Fourier decomposition to study the solution of a cubic: Lagrange transformed the roots x_1, x_2 x_3 into the resolvents:

$$r_1 = x_1 + x_2 + x_3$$

$$r_2 = x_1 + \zeta x_2 + \zeta^2 x_3$$

$$r_3 = x_1 + \zeta^2 x_2 + \zeta x_3$$

where ζ is a cubic root of unity, which is the DFT of order 3. A number of authors, notably Jean le Rond d'Alembert, and Carl Friedrich Gauss used trigonometric series to study the heat equation, but the breakthrough development was the 1807 paper Mémoire sur la propagation de la chaleur dans les corps solides by Joseph Fourier, whose crucial insight was to model all functions by trigonometric series, introducing the Fourier series.

Historians are divided as to how much to credit Lagrange and others for the development of Fourier theory: Daniel Bernoulli and Leonhard Euler had introduced trigonometric representations of functions, and Lagrange had given the Fourier series solution to the wave equation, so Fourier's contribution was mainly the bold claim that an arbitrary function could be represented by a Fourier series.

The subsequent development of the field is known as harmonic analysis, and is also an early instance of representation theory.

The first fast Fourier transform (FFT) algorithm for the DFT was discovered around 1805 by Carl Friedrich Gauss when interpolating measurements of the orbit of the asteroids Juno and Pallas, although that particular FFT algorithm is more often attributed to its modern rediscoverers Cooley and Tukey.

Interpretation in Terms of Time and Frequency

In signal processing, the Fourier transform often takes a time series or a function of continuous time, and maps it into a frequency spectrum. That is, it takes a function from the time domain into the frequency domain; it is a decomposition of a function into sinusoids of different frequencies; in the case of a Fourier series or discrete Fourier transform, the sinusoids are harmonics of the fundamental frequency of the function being analyzed.

When the function f is a function of time and represents a physical signal, the transform has a standard interpretation as the frequency spectrum of the signal. The magnitude of the resulting complex-valued function F at frequency ω represents the amplitude of a frequency component whose initial phase is given by the phase of F.

Fourier transforms are not limited to functions of time, and temporal frequencies. They can equally be applied to analyze spatial frequencies, and indeed for nearly any function domain. This justifies their use in such diverse branches as image processing, heat conduction, and automatic control.

REFERENCES

1. Rota, Gian-Carlo; Palombi, Fabrizio (1997), Indiscrete thoughts, Birkhäuser, ISBN 978-0-8176-3866-5, p. 11Neugebauer, Otto (1969) [1957], The Exact Sciences in Antiquity (2 ed.), Dover Publications, ISBN 978-0-486-22332-2
2. Brack-Bernsen, Lis; Brack, Matthias, Analyzing shell structure from Babylonian and modern times, arXiv:physics/0310126
3. Terras, Audrey (1999), Fourier analysis on finite groups and applications, Cambridge University Press, ISBN 978-0-521-45718-7, p. 30

4. Briggs, William L.; Henson, Van Emden (1995), The DFT : an owner's manual for the discrete Fourier transform, SIAM, ISBN 978-0-89871-342-8, p. 4
5. Briggs, William L.; Henson, Van Emden (1995), The DFT: an owner's manual for the discrete Fourier transform, SIAM, ISBN 978-0-89871-342-8, p. 2
6. Heideman, M.T.; Johnson, D.H.; Burrus, C.S. (1984). "Gauss and the history of the fast Fourier transform". IEEE ASSP Magazine. 1 (4): 14–21.
7. Knapp, Anthony W. (2006), Basic algebra, Springer, ISBN 978-0-8176-3248-9, p. 501
8. Narasimhan, T.N. (February 1999). "Fourier's heat conduction equation: History, influence, and connections" (PDF). Reviews of Geophysics. New York City, New York, United States: John Wiley & Sons. 37 (1): 151–172. doi:10.1029/1998RG900006. ISSN 1944-9208. OCLC 5156426043.
9. Terras, Audrey (1999), Fourier analysis on finite groups and applications, Cambridge University Press, ISBN 978-0-521-45718-7, p. 31

Index